The NASA STI Program Office... in Profile

Since its founding, NASA has been dedicated to the advancement of aeronautics and space science. The NASA Scientific and Technical Information (STI) Program Office plays a key part in helping NASA maintain this important role.

The NASA STI Program Office is operated by Langley Research Center, the lead center for NASA's scientific and technical information. The NASA STI Program Office provides access to the NASA STI Database, the largest collection of aeronautical and space science STI in the world. The Program Office is also NASA's institutional mechanism for disseminating the results of its research and development activities. These results are published by NASA in the NASA STI Report Series, which includes the following report types:

- TECHNICAL PUBLICATION. Reports of completed research or a major significant phase of research that present the results of NASA programs and include extensive data or theoretical analysis. Includes compilations of significant scientific and technical data and information deemed to be of continuing reference value. NASA's counterpart of peer-reviewed formal professional papers but has less stringent limitations on manuscript length and extent of graphic presentations.

- TECHNICAL MEMORANDUM. Scientific and technical findings that are preliminary or of specialized interest, e.g., quick release reports, working papers, and bibliographies that contain minimal annotation. Does not contain extensive analysis.

- CONTRACTOR REPORT. Scientific and technical findings by NASA-sponsored contractors and grantees.

- CONFERENCE PUBLICATION. Collected papers from scientific and technical conferences, symposia, seminars, or other meetings sponsored or cosponsored by NASA.

- SPECIAL PUBLICATION. Scientific, technical, or historical information from NASA programs, projects, and mission, often concerned with subjects having substantial public interest.

- TECHNICAL TRANSLATION. English-language translations of foreign scientific and technical material pertinent to NASA's mission.

Specialized services that complement the STI Program Office's diverse offerings include creating custom thesauri, building customized databases, organizing and publishing research results...even providing videos.

For more information about the NASA STI Program Office, see the following:

- Access the NASA STI Program Home Page at *http://www.sti.nasa.gov*

- E-mail your question via the Internet to help@sti.nasa.gov

- Fax your question to the NASA Access Help Desk at (301) 621–0134

- Telephone the NASA Access Help Desk at (301) 621–0390

- Write to:
  NASA Access Help Desk
  NASA Center for AeroSpace Information
  7121 Standard Drive
  Hanover, MD 21076–1320
  (301)621–0390

NASA/CR—2002–212050

Integrated In-Space Transportation Plan

B. Farris, B. Eberle, G. Woodcock, and B. Negast
Gray Research, Inc., Huntsville, Alabama

Prepared for Marshall Space Flight Center
under Contract H-32738D

National Aeronautics and
Space Administration

Marshall Space Flight Center • MSFC, Alabama 35812

October 2002

Available from:

NASA Center for AeroSpace Information	National Technical Information Service
7121 Standard Drive	5285 Port Royal Road
Hanover, MD 21076–1320	Springfield, VA 22161
(301) 621–0390	(703) 487–4650

 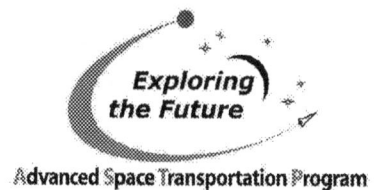

October 2002

To Whom It May Concern:

The Integrated In-Space Transportation Planning (IISTP) phase I activity was an assessment and prioritization of in-space propulsion technologies. This activity was conducted in 2001 by a NASA-wide team of over 100 engineers and scientists, resulting in a list of advanced in-space propulsion technologies benefiting multiple NASA Enterprises.

Over a 6-month period, the IISTP team evaluated primary propulsion systems intended to transport spacecraft from the launched condition to the destination and back, if required, for 28 potential missions. Seventeen propulsion technology architectures were evaluated, and priorities were assigned to the technologies according to their satisfaction of mission requirements, schedule, cost, and other selection criteria.

The enclosed report presents the prioritized set of advanced in-space propulsion technologies resulting from the IISTP activity and details of the supporting analysis.

Sincerely,
Les Johnson
In-Space Transportation Investment Area Manager
Advanced Space Transportation Program
Marshall Space Flight Center

Integrated In-Space Transportation Plan (IISTP) Phase I Final Report

14 September 2001

IISTP Phase I Final Report
September 14, 2001

PREFACE

The purpose of this report is to provide the reader with a readily accessible reference volume and history for the IISTP Phase I effort. This report was prepared by Gray Research, Inc, as a partial fulfillment of the Integrated Technology Assessment Center (ITAC) subcontract #4400037135 in support of the Integrated In-Space Transportation Plan (IISTP) Phase I effort within the In-Space Investment Area of the Advanced Space Transportation Program (ASTP) managed at Marshall Space Flight Center (MSFC) in Huntsville, Alabama. Much of the data used in the preparation of this report was taken from analyses, briefings and reports prepared by the vast number of dedicated engineers and scientists who participated in the IISTP Phase I effort. The opinions and ideas expressed in this report are solely those of the authors and do not necessarily reflect those of NASA in whole or in part.

EXECUTIVE OVERVIEW

Reaching the outer solar system is a struggle against time and distance. The most distant planets are 4.5 to 6 billion kilometers from the Sun and to reach them in any reasonable time requires much higher values of specific impulse than can be achieved with conventional chemical rockets. In addition, the few spacecraft that have reached beyond Jupiter have used gravity assist, mainly by Jupiter, that is only available for a few months' period every 13 or so years. This permits only very infrequent missions and mission planners are very reluctant to accept travel times greater than about ten years since this is about the maximum for which one can have a realistic program plan.

Advanced In-Space Propulsion (ISP) technologies will enable much more effective exploration of our Solar System and will permit mission designers to plan missions to "fly anytime, anywhere and complete a host of science objectives at the destinations" with greater reliability and safety. With a wide range of possible missions and candidate propulsion technologies with very diverse characteristics, the question of which technologies are "best" for future missions is a difficult one.

The primary focus of the IISTP Phase I efforts were to:
- Develop, iterate and baseline future NASA requirements for In-Space Transportation
- Define preliminary integrated architectures utilizing advanced ISP technologies
- Identify and prioritize ISP technologies

The primary efforts of the IISTP Phase I process was to:
- Address Customer defined missions, mission priorities, mission requirements and technology preferences.
- Provide a forum for Technologists to advocate and have sufficiently considered any ISP technology for any mission of interest defined by the customer.
- Perform Systems analyses of the customer defined prioritized mission set to the degree necessary to support evaluation and prioritization of each technology advocated by the technologists.
- Perform Cost analyses on each of the technologies that were determined by systems analyses to be viable candidates for the customer defined mission set.
- Integrate all customers, technologists, systems, cost, program and project inputs into the final IISTP Prioritized set of technologies.

The primary products of the IISTP Phase I effort were:
- Prioritized set of advanced ISP technologies that meet customer-provided requirements for customer prioritized mission sets
- Recommendations of relative technology payoffs to guide augmentation investments

The overall IISTP Phase I technology selection and prioritization process was accomplished in six steps:

1) Code Y, S and M Enterprises identified and prioritized a total of 28 missions. However, due to time constraints, only 9 missions were analyzed during Phase I.

2) The Enterprises developed a list of the figures of merit used to evaluate candidate advanced ISP technology systems for the missions identified.

3) Technologists identified 17 candidate ISP technologies for each of the 9 missions that could reasonably satisfy the mission requirements, objectives, cost and trip time objectives.

4) The Enterprises identified a list of 31 Figures of Merit (FOM) that were selected based on knowledge of the candidate ISP technologies and the missions for which the candidate technologies may be used. The Enterprises tailored the figures of merit for each of the missions through the use of weights. The weighting scale was adopted from the highly successful Kepner-Tregoe (K-T) method used throughout government and industry over the past forty years. The FOM weights were not disclosed to anyone on the ITSTP team until all scoring was completed to ensure the scoring teams scored each FOM independently without regard to their relative importance to one another.

5) Once mission analyses were completed, the scoring teams were provided with guidelines in the FOM Dictionary for scoring each of the candidate ISP technologies. Scoring guidelines were adopted from similar applications using Quality Function Deployment (QFD), widely used throughout the world since 1966. Other scoring methodologies such as Analytic Hierarchy Process (AHP) were considered but did not seem appropriate given the nature of the Phase I process. Out of almost 2500 scores assigned, less than ten were disputed. A sensitivity analysis, later applied, determined that none of the disputed scores had an effect on the overall final prioritization of ISP technologies.

6) Once all scoring was completed, the FOM category weights were applied to the scores and the cost-benefit assessment of each ISP technology for each mission was assessed by an independent multi-enterprise, multi-discipline team during a two-day workshop. ISP technologies were identified and prioritized during the workshop according to their relative payoff and their ability to perform and/or enable customer prioritized missions effectively and economically. The IISTP Phase I effort concluded with a consensus across NASA Programs, Projects, Technology Centers and Enterprises as to those technologies deserving consideration in future investment decisions.

TABLE OF CONTENTS

PREFACE	iii
EXECUTIVE OVERVIEW	v
1.0 Introduction	1
1.1 Overview	1
1.2 Outline	3
2.0 Background	5
2.1 Current Technology vs Advanced In-Space Propulsion Technologies	5
3.0 Organization	7
3.1 Mission Requirements Team (MRT)	8
3.2 Technology Team (TT)	9
3.3 Systems Team (ST)	10
3.4 Cost Team (CT)	11
3.5 IISTP Advisory Group (IAG)	12
4.0 Prioritization and Selection of ISP Technologies	13
4.1 Step One: Mission Identification and Prioritization	15
4.1.1 Neptune Orbiter	16
4.1.2 Titan Explorer	17
4.1.3 Europa Lander	18
4.1.4 Mars Sample Collection and Return	19
4.1.5 Interstellar Probe	20
4.1.6 Solar Polar Imager	21
4.1.7 Magnetospheric Constellation	22
4.1.8 Pole Sitter	23
4.1.9 HEDS Mars Piloted	24
4.2 Step Two: Figures of Merit (FOM) Development	25
4.3 Step Three: Candidate ISP Technology Identification	26
4.3.1 State-of-the-Art (SOA) Chemical	28
4.3.2 Advanced Chemical	29
4.3.2.1 Monopropellants	29
4.3.2.2 Bipropellants	29
4.3.2.3 Tripropellants	30
4.3.2.4 High Energy Density Matter (HEDM)	31
4.3.3 Nuclear Thermal (NTP)	32
4.3.4 Momentum Exchange (MX) Tether	33
4.3.4.1 Spinning Boost (Hohmann-type) Transfer	33
4.3.4.2 Swinging Boost Transfer	36

4.3.5 Electric Propulsion (EP)	37
4.3.5.1 Electrothermal Systems	39
4.3.5.2 Electrostatic Systems	39
4.3.5.2.1 Ion Thrusters	40
4.3.5.2.2 Hall Thrusters	42
4.3.5.3 Electromagnetic Propulsion	43
4.3.6 Solar Sails	44
4.3.7 Plasma Sails (M2P2)	46
4.3.8 Solar Thermal	49
4.3.9 NTP/NEP Bimodal	50
4.3.10 Aero-Capture	51
4.4 Step Four: Figures of Merit Weighting	53
4.5 Step Five: Evaluation and Scoring of Candidate ISP Technologies	55
4.5.1 Mission Analyses	55
4.5.2 Scoring Methodologies	55
4.5.3 Scoring Activities	56
4.5.3.1 Strawman Scoring	56
4.5.3.2 Data Presentation	57
4.5.3.3 Advocacy and/or Expert Input	57
4.5.3.4 Reaching Consensus	57
4.5.3.5 Recording of Results	57
4.5.3.6 IAG Review	57
4.5.4 Compilation and Presentation of Scoring Data	58
4.5.4.1 Normalized Weighted Score Sheets	58
4.5.4.2 Bar-Line Plots	60
4.5.4.3 FOM Category Weights	61
4.5.4.4 Effectiveness versus Cost	62
4.6 Step Six: Prioritization of ISP Technologies	66
4.6.1 Level III Decomposition	67
4.6.2 Level II Decomposition	68
4.6.3 Level I Decomposition	69
4.6.4 IISTP Technology Prioritization End-Product	70
5.0 Conclusions	71
6.0 Recommendations	73
7.0 Acronyms and Abbreviations	75
APPENDIX A IISTP Team Rosters	A-1
APPENDIX B Figures of Merit Dictionary	B-1
APPENDIX C Scores and Results	C-1
APPENDIX D Mission Analyses Results	D-1
APPENDIX E Technology Assessments	E-1
APPENDIX F Cost Team Report	F-1

FIGURES

Figure 1.1-1 IISTP Technology Development Strategy	1
Figure 4.0-1 IISTP Phase I Technology Prioritization and Selection Process	14
Figure 4.1.1-1 Neptune and Triton	16
Figure 4.1.2-1 Titan Explorer Mission	17
Figure 4.1.3-1 Europan Surface with Jupiter Rising	18
Figure 4.1.4-1 Martian Lander Return to Earth	19
Figure 4.1.5-1 Interstellar Probe Notional Flight	20
Figure 4.1.6-1 Solar Polar Imager's Orbit	21
Figure 4.1.7-1 Magnetospheric Constellation Deployment	22
Figure 4.1.8-1 Earth-Sun Interaction Graphic	23
Figure 4.1.8-2 Pole Sitter Satellite	23
Figure 4.1.9-1 Descent to Martian Surface	24
Figure 4.1.9-2 Martian Base and Rover	24
Figure 4.3.3-1 Schematic of a "Typical" Solid-Core Nuclear Rocket Engine	32
Figure 4.3.4.1-1 Artists Rendering of MX Orbital Capture	33
Figure 4.3.4.1-2 Orbits After Release In The "Spinning" Tether Boost Scenario	34
Figure 4.3.4.1-3 MX Tether Systems Reduce Launch Vehicle Size and Cost	35
Figure 4.3.4.2-1 MX Swinging Tether at Payload Release Point	36
Figure 4.3.5-1 Electric Propulsion Concepts	37
Figure 4.3.5-2 EP Systems Provide Significantly Reduce Trip Times	38
Figure 4.3.5.2-1 Basic Electrostatic Operation	40
Figure 4.3.5.2.1-1 Typical Ion Thruster	41
Figure 4.3.5.2.2-1 Schematic of a Hall Thruster	42
Figure 4.3.5.3-1 VASIMR Propulsion System	43
Figure 4.3.6-1 Solar Sail Concept	44
Figure 4.3.6-2 Solar Sail Dramatically Reduces Trip Times	45
Figure 4.3.7-1 Plasma Sail Inflation Demonstration in 2000	47
Figure 4.3.7-2 Spacecraft Using a Plasma Sail	47
Figure 4.3.7-3 M2P2 Sail Trip Time Reductions	48
Figure 4.3.8-1 Solar Thermal Propulsion Operating Principle	49
Figure 4.3.9-1 NTP/NEP Bimodal Operating Principle	50
Figure 4.3.10-1 Various Aerocapture Techniques	51
Figure 4.3.10-2 Aerocapture Reduces Requirements for Capture Maneuvers	52
Figure 4.5.4.1-1 Sample Scoring Sheet for Titan Explorer	59
Figure 4.5.4.2-1 Parallel Bar Chart for Titan Explorer	60
Figure 4.5.4.4-1 Cost/Effectiveness Scatter Chart - Titan Explorer Mission	63
Figure 4.5.4.4-2 Cost-Effectiveness Scatter Chart for Titan Explorer Mission with "Isos" overlays	64
Figure 4.5.4.4-3 Cost-Effectiveness Scatter Plot for NEP for All Missions	65
Figure 4.6-1 IISTP Technology Prioritization Workshop Process	66

TABLES

Table 4.1-1 IISTP Phase I Candidate Missions	15
Table 4.3-1 IISTP Phase I Mission/Technology Analyses Cross-Correlation	27
Table 4.3.2.2-1 Isp for a variety of Bipropellant Systems	30
Table 4.3.2.4-1 HEDM Isp Increases	31
Table 4.4-1 IISTP Phase I FOM Weights	54
Table 4.6.1-1 Level III Decomposition Results	67
Table 4.6.2-1 Level II Decomposition Results	68
Table 4.6.3-1 Level I Decomposition Results	69
Table 4.6.4-1 IISTP Phase I Consensus Results	70

1.0 Introduction

There is a significant interest within NASA for an increased investment in In-Space Propulsion (ISP) transportation technologies that:

Support multiple Enterprises within NASA,
Enable new missions,
Reduce mission costs and/or
Reduce travel time for planetary missions.

The In-Space Investment Area is responsible for implementing the Office of Space Science's (OSS) In-Space Propulsion (ISP) Program that supports the objectives of achieving a factor of 10 reduction in the cost of earth orbital transportation and a factor of 2 or 3 reduction in propulsion system mass and travel time for planetary missions within 15 years.

1.1 Overview

The Integrated, In-Space Transportation Plan (IISTP) technology development strategy, illustrated in Figure 1.1-1, focused on identification and prioritization of advanced ISP technologies that meet the objectives of OSS and the needs of the Agency as a whole.

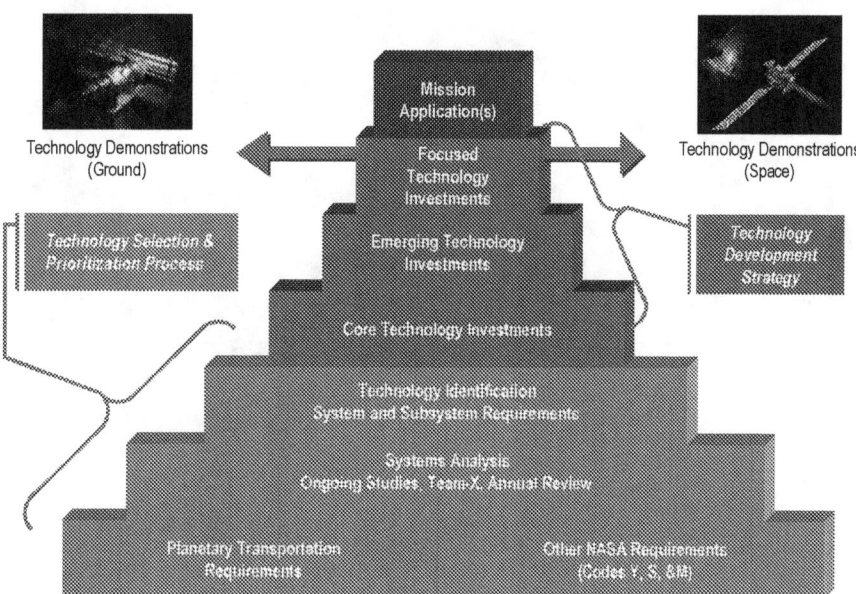

Figure 1.1-1. IISTP Technology Development Strategy

The IISTP Phase I effort focused on the ISP Technology Selection and Prioritization Process represented in Figure 1.0-1 as the bottom three tiers of the In-Space Investment Area Technology Development pyramid. The primary IISTP Phase I activities were:

- Develop, iterate and baseline future NASA requirements for In-Space Transportation

- Define preliminary integrated architectures utilizing advanced ISP technologies

- Identify and prioritize ISP technologies

- Assess program content, metrics and funding priorities and recommend options

The two primary products of the IISTP Phase I effort were:

1) Prioritized set of advanced ISP technologies that meet customer-provided requirements for customer prioritized mission sets,

2) Recommendations to OSS management and OMB of relative technology payoffs of selected ISP technologies to guide investments.

1.2 Outline

This report is organized into the following six sections:

Section 1.0 Introduction- Overview of IISTP effort and purpose and organization of the for the report.

Section 2.0 Background- Discussion of why the IISTP effort is important to the future exploration of space.

Section 3.0 Organization- Identification of the IISTP Phase I teams and their respective roles and responsibilities.

Section 4.0 Prioritization and Selection of ISP Technologies- An in-depth look at the IISTP Phase I process that was accomplished in the following six steps:

    Section 4.1 Step One: Mission Identification and Prioritization

    Section 4.2 Step Two: Figures of Merit Development

    Section 4.3 Step Three: Candidate ISP Technology Identification

    Section 4.4 Step Four: Figures of Merit Weighting

    Section 4.5 Step Five: Evaluation and Scoring of Candidates

    Section 4.6 Step Six: Prioritization of ISP Technologies

Section 5.0 Conclusions- Brief summary and concluding remarks

Section 6.0 Recommendations- Discussion of lessons learned and recommendations for follow-on efforts

Section 7.0 Acronyms

Supplemental reference material was organized in the following Appendices:

Appendix A IISTP Team Rosters- Identification by name of the members on each of the five Phase I teams

Appendix B Figures of Merit Dictionary- The Figures of Merit Dictionary (Rev E) used during the Phase I effort

Appendix C Scores and Results- A compilation of all bar-line and scatter plots generated from the Phase I scoring data

Appendix D Mission Overviews- Supplemental data on the missions analyzed.

Appendix E Technology Assessments- Compilation of the results of independent assessments performed during Phase I

Appendix F Cost Team Report- The Cost Analysis Report written by the Cost Team during Phase I

2.0 Background

Advanced In-Space Propulsion (ISP) technologies will enable much more effective exploration of our Solar System. ISP technologies will permit mission designers to plan missions to "fly anytime, anywhere and complete a host of science objectives at the destination" with greater reliability and safety.

<u>Fly anytime, arrive sooner:</u> Advanced ISP technologies will reduce reliance on planetary flyby gravity-assist maneuvers to reach the destination, i.e. launches need not wait on infrequent planetary alignments needed for flyby assists. In addition, advanced ISP systems will enable significantly reduced trip times as illustrated in Figure 2.0-1.

<u>Freedom from Constraints</u>: Advanced ISP technologies will enable mission success to be predicated on satisfaction of science objectives rather than the need to overcome transportation constraints. In addition, at the destination, more complex science gathering missions can be accomplished with superior maneuverability, ascent/descent and station keeping capabilities.

<u>Reduced Cost</u>: Since advanced ISP technologies will dramatically reduce overall mission timelines, operational costs can be significantly reduced. Smaller launch vehicles can be used in most cases. In addition, extended capabilities of advanced ISP systems will radically reduce the number of missions required to accomplish the same science objectives.

2.1 Current Technology versus Advanced In-Space Propulsion Technologies

With the exception of electric propulsion systems used for commercial communications satellite station-keeping, all of the rocket engines in use today are chemical rockets; that is, they obtain the energy needed to generate thrust by combining reactive chemicals to create a hot gas that is expanded to produce thrust. A significant limitation of chemical propulsion is that it has a relatively low specific impulse (thrust per unit of mass flow rate of propellant). Numerous concepts for advanced propulsion technologies with significantly higher values of specific impulse have been developed over the past fifty years. However, they generally have very small values of thrust. For launch from the surface of the earth to low earth orbit, large thrust is required to overcome the effect of the earth's gravity. For this reason, chemical propulsion has remained as the primary propulsion technology because it is the only propulsion technology capable of producing the magnitude of thrust necessary to overcome the effect of gravity.

Once earth orbit is achieved, high thrust is no longer required. Low thrust technologies can be used if they can be operated for long durations. Several of these technologies offer specific impulse that is significantly higher than that achievable with chemical propulsion. The advantage of high specific impulse in achieving high flight speeds is expressed in the conventional rocket equation:

$$\Delta V = Isp \ln(mi/mf)$$

Where ΔV = Change in vehicle velocity imparted by propulsion system
  Isp = Specific impulse
  mi = Initial mass of rocket stage (including payload)
  mf = Final mass of rocket stage (including payload)

This equation shows that the velocity imparted by a rocket stage is directly proportional to specific impulse.

Reaching the outer planets requires traversing of enormous distances. The most distant planets are 4.5 to 6 billion kilometers from the sun. To reach them at all, considering the strength of the sun's gravity field, requires high velocities. To reach them in reasonable time requires much higher velocities. The few spacecraft that have reached beyond Jupiter have used gravity assist, mainly by Jupiter, to attain these velocities. However, a Jupiter gravity assist for reaching a particular destination is only available for a few months every 13 or so years. This permits only very infrequent missions.

The exploration of the outer planets clearly requires development of advanced propulsion concepts for which an impetus for development has not previously existed. They are required to decrease trip time, increase payload mass fraction, and enable missions that are not feasible with chemical propulsion. The existence of many concepts requires the careful selection of a few concepts for development to flight status. This selection must match the characteristics of the propulsion technology with the requirements of a diverse set of anticipated space missions, particularly those to the outer planets and beyond.

With a wide range of possible missions and candidate propulsion technologies with very diverse characteristics, the question of which technologies are "best" for future missions is a difficult one. The IISTP study is a rational process to select and prioritize propulsion technologies for development to flight status for anticipated future space missions, particularly those to the outer planets and beyond.

3.0 Organization

The IISTP Phase I organization was created to ensure:

1) <u>Customer</u> defined missions, mission priorities, mission requirements, figures of merit, weights and technology preferences could be identified and adequately captured during the IISTP process.

2) <u>Technologists</u> had a forum to advocate any ISP technology for any mission of interest defined by the customer.

3) <u>Systems</u> analyses were performed for the customer defined prioritized mission set to the degree necessary to support evaluation and prioritization of each technology advocated by the technologists.

4) <u>Cost</u> analyses were performed on each of the technologies that were determined by systems analyses to be viable candidates for the customer defined mission set.

5) <u>Integration</u> of all customers, technologists, systems, cost, program and project inputs into the IISTP prioritization process.

Five teams were formed, each with its own roles and responsibilities to facilitate satisfaction of these organizational objectives. The five teams by name were the:

1) Mission Requirements Team (MRT)

2) Technology Team (TT)

3) Systems Team (ST)

4) Cost Team (CT)

5) IISTP Advisory Group (IAG)

Team rosters are given in Appendix A. The specific roles and responsibilities for each of these teams are discussed in the subsections that follow.

3.1 Mission Requirements Team (MRT)

The mission requirements team (MRT) was comprised of one or more representatives from each of the Space Science (Code S), Earth Science (Code Y) and Human Exploration and Development of Space (Code M) Enterprises. MRT members represented their respective Enterprises throughout the IISTP process.

Individual MRT members were primarily responsible for:

- Identification of a prioritized set of missions for their respective Enterprises

- Identification of the figures of merit to be used in evaluation of candidate ISP technologies

- Determination of figures of merit weightings used in evaluation of candidate ISP technologies

- Participation in weekly IAG telecons

- Participation in the IISTP Prioritization Workshop

3.2 Technology Team (TT)

The Technology Team (TT) was comprised of a representative(s) for each ASTP technology element and appropriate research areas which included electric propulsion, sails, fission, tethers, aero-assist, in-situ propellant production, advanced chemical, lightweight components, cryogenic fluid management and solar thermal propulsion. A single point of contact (POC) was named to represent and lead the TT team effort.

Individual TT members were primarily responsible for:

- Participation in ST telecons, reviews and analysis meetings

- Identification of candidate ISP technologies in support of customer prioritized mission set

- Providing the ST with candidate technology characteristics to the extent necessary to support systems analyses and evaluations of the technologies for each of the missions analyzed

- Evaluation and scoring of each technology used in mission analyses against the reliability/safety and schedule related figures of merit

- Participation in ST and CT telecons, reviews and analysis meetings

- Participation in weekly IAG telecons by TT lead.

- Participation in the IISTP prioritization workshop by TT lead

3.3 Systems Team (ST)

The Systems Team (ST) included representatives from the systems organizations at MSFC, GRC, JSC, ITAC, JPL, JSC, and LaRC. A single POC was named to represent and lead the ST team effort.

Individual ST members were primarily responsible for:

- Development of systems concepts and architectures for each of the missions prioritized by the MRT

- Performing systems analyses of the MRT prioritized missions using each of the candidate ISP technologies identified by the TT

- Evaluation and scoring of each technology used in the mission analyses against the performance and technical related figures of merit

- Participation in TT and CT telecons, reviews and analyses meetings

- Participation in weekly IAG telecons by ST lead

- Participation in the IISTP Prioritization Workshop by ST Lead

3.4 Cost Team (CT)

The Cost Team (CT) was comprised of MSFC and support contractor cost analysts as well as a liaison from the ST. A single POC was named to represent and lead the CT effort.

The CT was primarily responsible for:

- Performing development cost analyses of the ISP candidate technologies selected for analyses for the MRT prioritized set of missions

- Evaluation and scoring of each technology used in the mission analyses against the cost related figures of merit

- Participation in ST and TT team telecons, reviews and analyses meetings

- Participation in weekly IAG telecons by CT lead

- Participation in the IISTP Prioritization workshop by CT lead

3.5   IISTP Advisory Group (IAG)

The IISTP Advisory Group (IAG) was comprised of the three POCs from the Code Y, S, and M enterprises (MRT POCs), the three leads from the Technology, Systems and Cost Teams, the two In-Space Investment Area Project Managers, along with senior advisors from GRC, and JPL. The In-Space Investment Area Program Manager chaired the IAG.

The IAG was primarily responsible for:

- Oversight of all IISTP activities

- Integration of TT, ST and CT activities

- Development and maintenance of the FOM Dictionary

- Consolidation and maintenance of FOM weights

- Development, maintenance and implementation of an IISTP Technology Prioritization Process

- Participation in and conduct of weekly IAG telecons

- Participation in and conduct of the IISTP Prioritization Workshop

4.0 Prioritization and Selection of ISP Technologies.

The overall IISTP Phase I technology selection and prioritization process was accomplished in six steps as illustrated in Figure 4.0-1.

In Step 1, the MRT developed a prioritized list of missions and their respective mission requirements to be addressed in the IISTP Phase I effort.

In Step 2, the MRT developed a list of the figures of merit that could be used to evaluate candidate advanced ISP technology systems for the missions identified.

In Step 3, the TT identified candidate ISP technologies for each mission that could reasonably satisfy the mission requirements, objectives, cost and trip time objectives.

In Step 4, the MRT tailored the figures of merit for each of the mission categories through the use of weights.

In Step 5, the ST, TT and CT evaluated how well each of the candidate ISP technologies satisfied each one of the applicable figures of merit. The ST, TT and CT used the scoring convention and scoring guidelines given in the Figure of Merit Dictionary developed by the IAG.

Steps 3 through 5 were repeated for each of the nine missions analyzed during the IISTP Phase I effort.

In Step 6, the IAG applied the mission category figures of merit weights and generated plots of the normalized scores. The scoring data was reviewed in a two-day workshop to identify and develop a prioritized set of ISP technologies to be used to guide investment decisions.

Each of these six steps is discussed in detail in the subsections that follow.

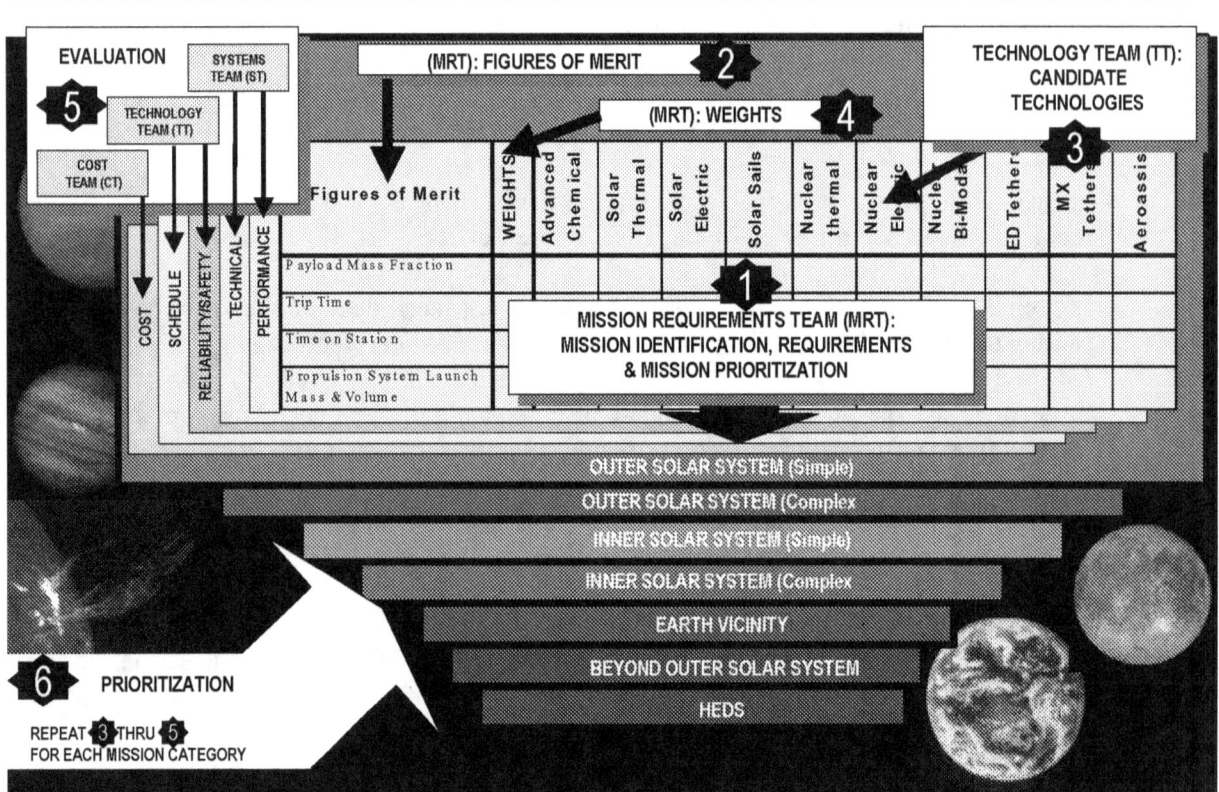

Figure 4.0-1. IISTP Phase I Technology Prioritization and Selection Process

4.1 Step One: Mission Identification and Prioritization

The MRT identified a total of 28 missions that were of interest to the Code Y, S, and M Enterprises. The 28 missions were allocated to one of nine different mission categories according to mission destination and need for propulsion at the destination (See Table 4.1-1). To ensure the highest priority missions were analyzed first, the Code Y, S and M POCs prioritized their respective missions within each mission category. Accordingly, the nine missions that were analyzed during IISTP Phase I are denoted by italics in Table 4.1-1. For each mission analyzed, the appropriate Enterprise POC provided the top-level mission requirements that were documented and maintained in the IITSP Requirements Document. An overview of each of these nine missions is provided in the sections that follow.

	Subsection	Category 1: Earth vicinity, low to moderate Delta V
1		Geospace Electrodynamic Connections (GEC)
2		EREMF (Leonardo)
3		Nat SAR
4		LEO SAR
5	4.1.7	*Magnetospheric Constellation*
6		Ionospheric Mappers
		Category 2: Inner solar system, simple profile, moderate Delta V
7		Space Interferometry Mission (SIM)
8		Starlight ST-3
		Category 3: Inner solar system, sample return
9		Comet Nucleus Sample Return (CNSR)
10	4.1.4	*Mars Sample Return*
		Category 4: Inner solar system, complex profile, moderate to high Delta V
		(Lagrange point missions will be considered as complex due to the sensitivity of the trajectory to perturbations. **= E-S; E-M Li point missions)
11		EASI**
12	4.1.8	*Pole Sitter***
13		Sub L1 point mission**
14		Solar Sentinels**
15	4.1.6	*Solar Polar Imager*
16		NGST**
17		Terrestrial Planet Finder**
18		Outer Zodiacal Transfer**
		Category 5: Outer solar system, simple profile, high Delta V
19		Outer Zodiacal Transfer
		Category 6: Outer solar system, complex profile, incl. propulsion in the outer solar system
20	4.1.2	*Titan Organics Explorer Orbiter & Lander***
21	4.1.1	*Neptune Orbiter*
22	4.1.3	*Europa Lander*
23		Solar Probe
		Category 7: Beyond outer solar system
24	4.1.5	*Interstellar Probe***
		Category 8: HEDS lunar, cislunar, & Earth vicinity
25		Moon & Earth-Moon Libration Points
26		Sun-Earth Libration Points
		Category 9: HEDS Asteroids / Mars vicinity
27		Near Earth Asteroids
28	4.1.9	*Mars Cargo & Piloted*

NOTE: Missions indicated by red italics were analyzed during IISTP Phase I

Table 4.1-1. IISTP Phase I Candidate Missions

4.1.1 Neptune Orbiter

The Neptune orbiter mission is designed to provide valuable insight into the eighth planet in our solar system and its largest moon Triton. After a 10-year flight, the Neptune Orbiter will spend 2 to 4 years on station while utilizing advanced communication techniques to relay valuable information about Neptune's atmospheric and magnetospheric properties. The orbiter will also perform multiple flybys of Triton providing physical and atmospheric information on this satellite. Since Triton is thought to be a Kuiper Belt object captured by Neptune, insight into the origins of our solar system and its continued development are a primary mission goal. Figure 4.1.1-1 is a colorized collage of Neptune and Triton.

Neptune is about 30 AU from Earth. Achieving a 10-year trip requires high velocity, which stresses the transportation system. Also, the velocity required results in a high encounter velocity at Neptune, about 12 km/sec. Capture requires either significant propulsive delta V or an aerocapture device capable of an entry speed about 30 km/sec.

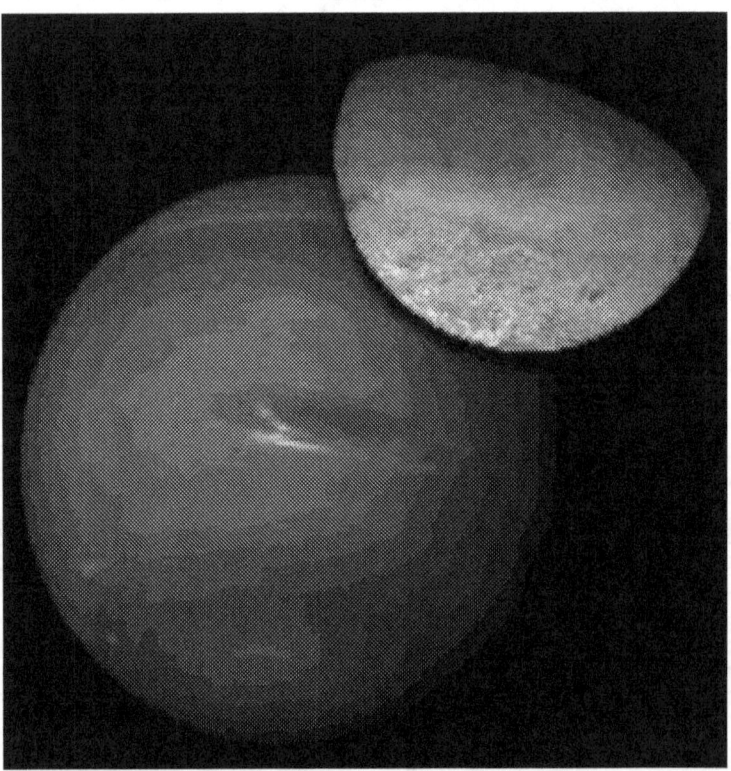

Figure 4.1.1-1 Neptune and Triton

[1]http://www.jpl.nasa.gov/adv_tech/ballutes/misn_neptune.htm

4.1.2 Titan Explorer

The Titan explorer mission is designed for orbital and surface analysis of Saturn's largest moon. After almost a 10-year flight, the spacecraft will make multiple orbits of Titan before deploying an advanced robotic lander with a mini chemistry lab to the surface. This lander will have the capability to change locations via ground and/or flight to collect and analyze surface samples at various locations on the moon. A variety of measurements and analysis will be performed in orbit and on the surface with the results sent back to earth via an advanced communications suite. Figure 4.1.2-1 shows a concept of the Titan Explorer mission.

Figure 4.1.2-1 Titan Explorer Mission

[2]http://www.jpl.nasa.gov/adv_tech/ballutes/misn_titan.htm

IISTP Phase I Final Report
September 14, 2001

4.1.3 Europa Lander

The Europa Lander mission concept is such that after a 3-year flight to Jupiter's fourth largest moon, the vehicle will spend several weeks in orbit around Europa before sending a robotic landing craft to the surface. The intention is to bury the craft just below the surface to protect it from radiation hazards and increase its ability to take seismic measurements. The planned 10 days of surface/subsurface analysis should provide valuable insight into the ice sheets and topography of this moon. Figure 4.1.3-1 shows an artist's rendering of the Europan surface with Jupiter in the background.

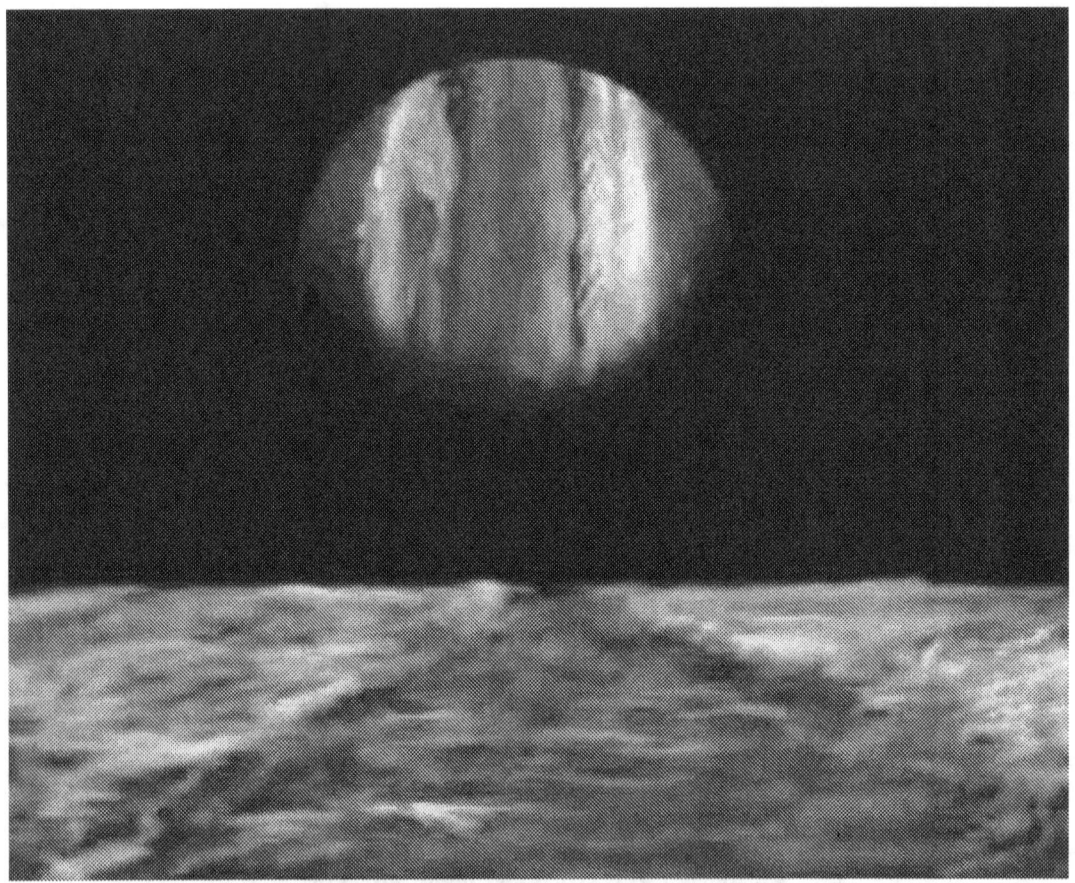

Figure 4.1.3-1 Europan Surface with Jupiter Rising

[3]http://sse.jpl.nasa.gov/site/missions/B/europa_lander_network.html

4.1.4 Mars Sample Collection and Return

The Mars Sample Return mission is part of NASA's continued exploration of the Red Planet. The spacecraft will fly to Mars, land, and return with soil, rock, and atmospheric samples. Robotics will be utilized to the maximum extend possible to allow samples to be collected from various locations around the landing sight. This mission could serve as a precursor to a manned flight to Mars, which may take place later in the decade. Figure 4.1.4-1 is a conceptualized version of the Martian lander blasting off to return to Earth.

This is a complex mission requiring Earth launch and transfer to Mars, capture of a spacecraft into Mars orbit, landing on Mars, launch of the sample carrier from Mars, rendezvous and sample transfer to the orbiting craft, return to Earth orbit, and sample return to Earth's surface. Strict contamination protection rules for contamination of Mars and back-contamination of Earth apply to the mission design. Some mission architectures use variations on the profile described here, such as direct launch from Mars' surface to Earth, bypassing rendezvous in Mars orbit. None of the individual delta Vs are especially high, but the total delta V, considering all profile elements, stresses the in-space transportation system and places a premium on performance and reducing inert mass.

Figure 4.1.4-1 Martian Lander Return to Earth

[4] http://www.jpl.nasa.gov/adv_tech/ballutes/misn_mars.htm

4.1.5 Interstellar Probe

The Interstellar Probe is intended to analyze the interstellar medium, the space between the stars of our galaxy. Our Sun's heliosphere shields us from the interstellar medium, so very little is known about the vast areas of space between stars. The Interstellar Probe will utilize advanced propulsion technology to quickly leave the influence of our own Sun's heliosphere and explore the area of space adjacent to, but outside our Sun's influence. As it travels to the edges of the heliosphere, it will also take data on heliosphere-interstellar medium interactions. Figure 4.1.5-1 is an artist's rendition of the Interstellar Probe's flight path.

The nominal performance target for this mission is to reach a distance of 200 AU in twenty years or less. This requires a delta V beyond Earth escape of about 35 km/s, assuming it is delivered in a few months or less. For longer thrusting periods, the delta V goes up to 40 to 50 km/s. Only the highest performance in-space propulsion systems are practical for this very demanding mission.

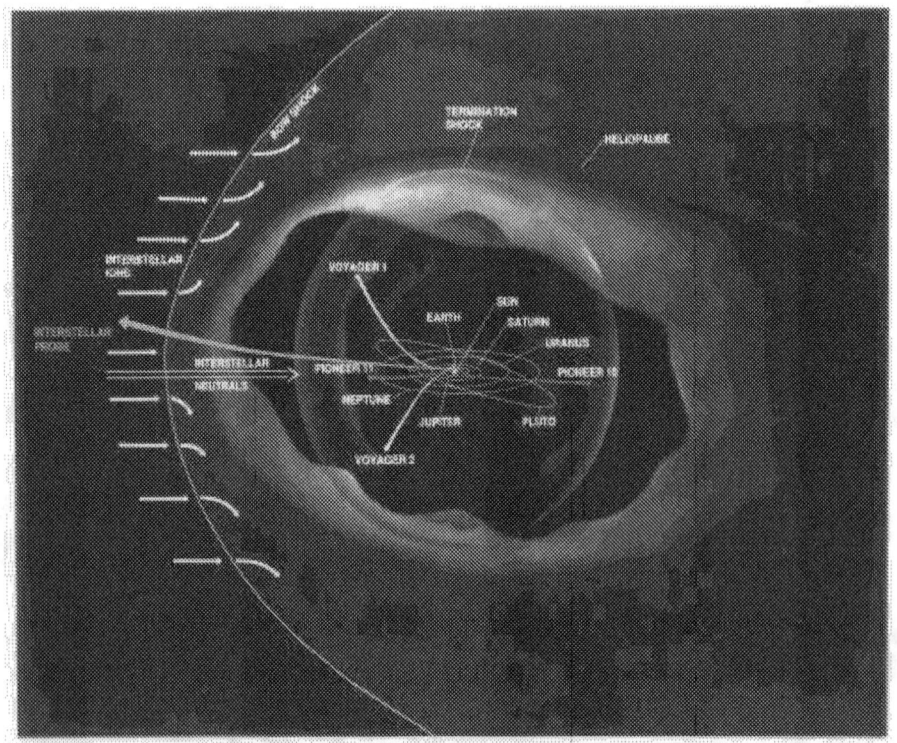

Figure 4.1.5-1 Interstellar Probe Notional Flight

[5] http://science.nasa.gov/ssl/pad/solar/suess/Interstellar_Probe/ISP-Intro.html

4.1.6 Solar Polar Imager

To fully understand the structure of the solar corona and to obtain a three-dimensional view of coronal mass ejections, we will need observations from above the Sun's poles to complement data obtained from the ecliptic plane. Solar observations could start as soon as two years after launch with a planned duration of 3 years. Viewing the Sun and inner heliosphere from a high-latitude perspective could be achieved by a solar polar imager in a Sun-centered orbit about one half the size of Earth's orbit, perpendicular to the ecliptic. Figure 4.1.6-1 shows the planned orbit of the Solar Polar Imager.

This mission requires a heliocentric plane change to go from a near-ecliptic path to one inclined 45 deg. or more to the ecliptic. It must also go close to the Sun, to about half Earth's distance. The delta V requirement is large, and favors high-Isp systems or those that derive thrust from solar interactions, such as solar sails.

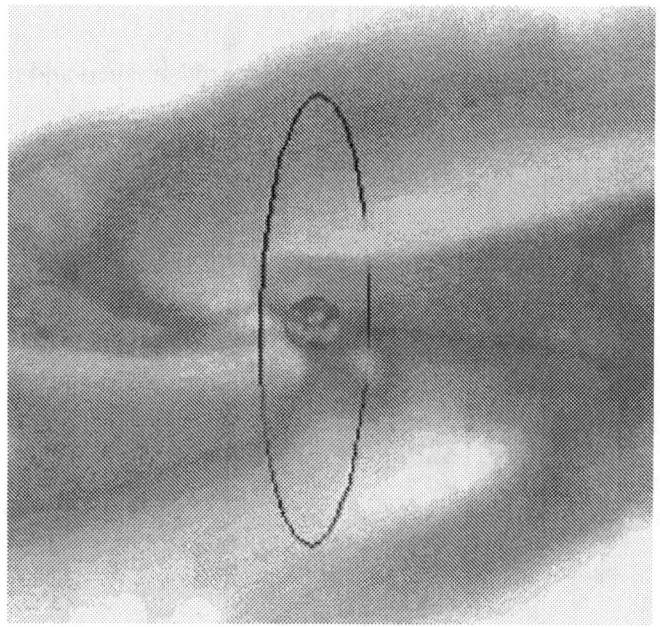

Figure 4.1.6-1 Solar Polar Imager's Orbit

[6]http://umbra.nascom.nasa.gov/spd/secr/missions/polarimg.html

4.1.7 Magnetospheric Constellation

The Magnetospheric Constellation mission intends to study the magnetotail of the Earth. The magnetotail is the large magnetic field trailing Earth's orbit around the Sun. A constellation of 50-100 nano-satellites will be deployed in elliptical orbits (and possibly orbital planes) around the Earth. These orbits have the same apogee at approximately 3 Earth Radii (R_E), with varying perigees from 7 to 40 R_E, creating a distributed network of space weather observatories. Mission delta V is not high but many small impulses are required. Figure 4.1.7-1 shows deployment of the nanosatellites in Earth's magnetotail. The primary objectives of this mission are:

- Determine the *equilibria* of the magnetotail
- Understand the *responses* of the magnetotail to the solar wind
- Reveal the *instabilities* of the magnetotail
- Elliptical orbits with dense sampling from 7 - 40 R_E with a resolution of 1-2 R_E
- Measure magnetic and plasma scalar and vector fields
- Track propagating fronts and disturbances as they are launched and travel in the magnetotail
- Develop synoptic maps of plasma flows into and away from magnetotail particle acceleration regions

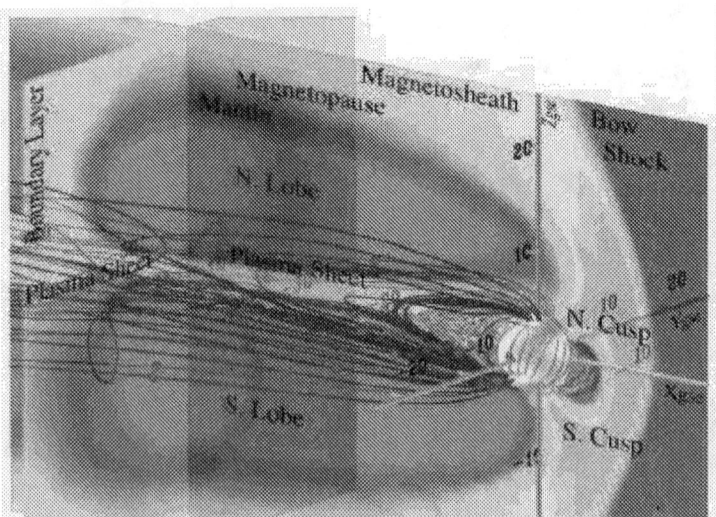

Figure 4.1.7-1 Magnetospheric Constellation Deployment

[7] http://stp.gsfc.nasa.gov/missions/mc/mc.htm

4.1.8 Pole Sitter

This is an Earth Science (Code Y) Mission with cooperation between NASA, the NOAA, and several other agencies to study sun-earth interactions causing the solar weather. These satellites will hopefully be part of a larger constellation around and between the Earth and the Sun in order to completely study all aspects of the sun's influence on our planet as shown in Figures 4.1.8-1 and 4.1.8-2. Two pole sitter satellites will be placed in orbits above each of the Earth's poles at a distance of approximately 60 R_E. Since these are not stable orbits, constant thrusting via advanced propulsion technology will be necessary to keep the satellites on station for the duration of the mission.

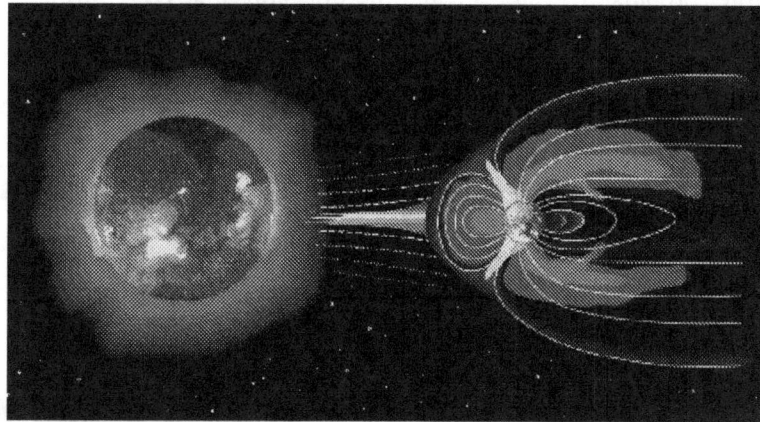

Figure 4.1.8-1 Earth-Sun Interaction Graphic

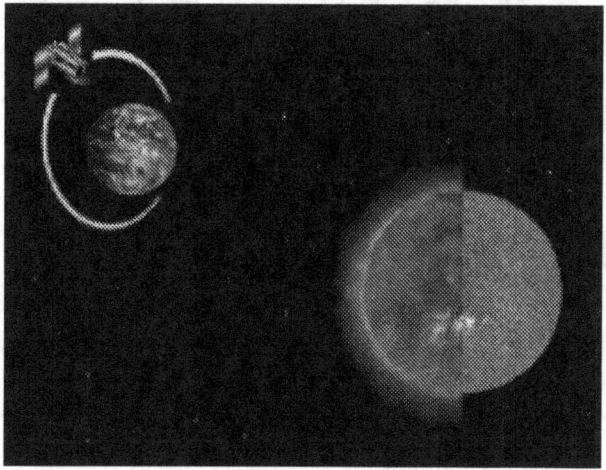

Figure 4.1.8-2 Pole Sitter Satellite

[8] http://lws.gsfc.nasa.gov/lws_resources_imagegallery.htm

4.1.9 HEDS Mars Piloted

A manned trip to Mars is the natural extension of continued exploration of our solar system. Mission objectives include developing a better understanding of Mars both currently and historically and to demonstrate the feasibility of future longer-term Mars exploration and/or colonization. Manned launches would likely be combined with cargo launches to provide backup equipment and supplies for the first and future manned exploration missions. Figure 4.1.9-1 shows a conceptual Martian lander descending to the Martian surface. Figure 4.1.9-2 shows the deployed lander and rover ready to explore the surface of Mars. Mission payloads are large, 10s to 100s tons; mission delta Vs can be high, depending on mission profile. This mission requires high performance and much larger propulsion systems than other missions analyzed during IISTP Phase I.

Figure 4.1.9-1 Descent to Martian Surface

Figure 4.1.9-2 Martian Base and Rover

[9] http://nssdc.gsfc.nasa.gov/planetary/mars/mars_crew.html

4.2 Step Two: Figures of Merit (FOM) Development

Candidate ISP technologies that satisfied the requirements of each the nine missions selected were analyzed. However, additional criteria over and above the mission requirements were needed to evaluate the relative merits of each of these candidate technologies. The MRT identified a list of 31 Figures of Merit (FOM) that were selected based on knowledge of the candidate ISP technologies and the mission categories for which the candidate technologies may be used. The MRT supported the development of a FOM Dictionary (See Appendix B) that was used to:

- Define each FOM

- Provide guidance for scoring the candidate ISP technologies

- Identify the scoring responsibilities

These 31 FOM were grouped into 6 categories according to *function*:

1) Performance – These criteria, scored by the ST, are directly related to how well each candidate ISP technology performed the mission.

2) Technical – These criteria, scored by the ST and the TT, are measures of technical robustness associated with each candidate ISP technology.

3) Reliability/Safety – These criteria, scored by the TT, address the inherent reliability of the technology and the relative ease to achieve required safety margins.

4) Cost - These criteria, scored by the CT, included measures of recurring, non-recurring, operational and developmental costs.

5) Applicability – These criteria were to be used to access the inherent applicability, adaptability, flexibility, scalability, and evolutionary capability of each technology. The IAG decided to defer consideration of these FOMs until the IISTP Phase I Workshop.

6) Schedule – These criteria, scored by the TT, were used to assess the maturity of the technology and the risks associated with development schedules.

Since consideration of the five FOM associated with the Applicability category was deferred until the IISTP Phase I workshop, initial scoring and evaluations were accomplished only on the remaining 5 categories and the associated 26 FOM.

4.3 Step Three: Candidate ISP Technology Identification

For each mission to be analyzed, the technology advocates "lobbied" for their respective technologies. Based on how well each technology satisfied the mission requirements, objectives, architecture, cost and trip time objectives, the TT worked with the ST and the MRT to identify those candidates that could most reasonably accomplish each mission. This was an iterative process and resulted in some missions with nearly 20 candidate ISP "systems" and others with fewer than 5. An ISP "system" may involve a combination of ISP technologies. The 25 ISP "systems" analyzed during the IISTP Phase I effort, were comprised of 17 unique ISP technologies that are shown in Table 4.3-1 cross-referenced to the 9 missions analyzed.

The Neptune Orbiter was the first mission analyzed. All technology candidates were evaluated for this mission in order to develop an understanding of technology applicability. Scoring proved to be very time-consuming, thus it was important to "thin out" the field of candidates in order to have time to score more missions.

For example, 7 of the 17 ISP technologies were various forms of electric propulsion (EP) systems. The differences among the EP systems had to do with the power system used, thruster type and size. Upon analyzing and scoring the Neptune Orbiter mission, the scoring teams determined that the variations among the different EP thrusters for the different planetary missions could be reasonably ascertained using the Neptune Orbiter results. Therefore it was not necessary to score all of the EP thruster options for other planetary missions.

The Titan Explorer mission analyses were performed by JPL's Team X. They chose not to evaluate certain technologies, either because they had insufficient data, or because the results of Neptune Orbiter scoring indicated certain technologies would not be effective for the mission.

Europa Lander was evaluated for all technologies except the EP thruster variations, solar thermal and NTP/NEP hybrid. Neptune Orbiter results confirmed that solar thermal propulsion has too little Isp to be competitive for these planetary missions. The NTP/NEP hybrid was defined as a system where an NEP vehicle is booster by an NTP. (The two functions are combined in a single engine in the NTP bimodal.) This option appeared effective only for extremely demanding missions, and was further evaluated only for the interstellar probe and the HEDS Mars piloted mission.

The Mars Sample Return mission was the last mission analyzed. There was not sufficient time for adequate mission analysis and scoring to be performed on many of the ISP systems. Since SEP Ion systems were the common thread running across 8 of the 9 missions it was used as a reference system.

For the Interstellar Probe, Solar Polar Imager, Magnetospheric Constellation, Pole Sitter and HEDS Mars Piloted missions, all technologies capable of performing these demanding missions or offer unique benefits were analyzed and scored (exempting variations in EP systems as discussed earlier)

An overview of each of these technologies is provided in the subsections that follow according to the reference subsections given in Table 4.3-1

Ref. No.		Reference Subsection	Neptune Orbiter	Titan Explorer	Europa Lander	Mars Sample Return	Interstellar Probe	Solar Polar Imager	Magneto-spheric Con-stellation	Pole Sitter	HEDS Mars Piloted	TOTAL
	Category		6	6	6	2	7	3	1	3	9	
2	SOA Chemical	4.3.1	YES	YES	YES	YES				YES		5
3, 24	Advanced Chemical	4.3.2	YES		YES	YES						3
4	Nuclear Thermal (NTP)	4.3.3	YES	YES	YES						YES	4
5	NTP Bimodal		YES		YES	YES					YES	4
6-7	MX Tether	4.3.4	YES		YES							2
8-9	Solar Electric (Hall)	4.3.5.2.2	YES							YES		2
10-11, 23, 25	Solar Electric (Ion)	4.3.5.2.1	5/10 kW	5/10 kW	5/10 kW	YES		5/10 kW	5/10 kW	5/10 kW NSTAR	100 kW	8
12	Nuclear Electric (Hall)	4.3.5.2.2	YES				YES					2
13	Nuclear Electric (Ion)	4.3.5.2.1	YES	YES	YES	YES	YES	YES			YES	7
14	Nuclear Electric (VaSIMR)	4.3.5.3	YES								YES	2
15	Nuclear Electric (MPD)	4.3.5.3	YES								YES	2
16	Solar Sails	4.3.6	YES	YES	YES		YES	YES	YES	YES		7
17-18	M2P2	4.3.7	YES	YES	YES	YES	YES					5
19	Radio-Isotope Electric	4.3.5.2.1	YES									1
20	Solar Thermal	4.3.8	YES							YES		2
21	NTP/NEP Hybrid	4.3.9					YES				YES	2
n/a	Aero-Capture	4.3.10	YES	YES	YES	YES						4
	TOTAL		16	7	10	7	5	3	4	3	7	

Table 4.3-1. IISTP Phase I Mission/Technology Analyses Cross-Correlation

4.3.1 State-of-the Art (SOA) Chemical

Chemical propulsion has historically been the primary means for transportation of payloads in space. Because chemical propulsion systems can generate the very large thrust required to overcome the effect of earth's gravity, they remain the preferred choice for launch to low earth orbit. Chemical rockets have been used for in-space transportation because they are understood well and are relatively cheap to develop. However, inherent performance limitations associated with chemical propulsion severely restrict the types of missions and destinations that can be achieved in a reasonable time, especially for destinations that are far from earth. Chemical propulsion is energy limited since the quantity of energy released during the combustion process is fixed by propellant chemistry. This limited quantity of energy limits specific impulse (thrust per unit of mass flow rate of propellant). This causes chemical propellant mass fractions to be high, while the payload fractions are low, resulting in expensive, inefficient missions. For launch from earth, these limitations are overcome by multiple stage launch systems.

Specific impulse for liquid propellant systems is limited to about 450 lbf-s/lbm. Unfortunately, these values of specific impulse are only possible for cryogenic propellants, resulting in difficult propellant handling issues, both on the ground and in space. Nitrogen tetroxide/monomethyl-hydrazine (NTO/MMH) propellants are storable on earth and in space, but have values of specific impulse of about 317 lbf-s/lbm. NTO/MMH has the advantage of being hypergolic, which means that the propellants react on contact, eliminating the need for any ignition system. Solid rockets have lower values of specific impulse (237 lbf-s/lbm), but they have higher values of density impulse (delivered impulse per unit of volume of propellant). Solid rockets are most often beneficial when reductions in aerodynamic drag are important, since tankage volume and thereby frontal area can be minimized. Monopropellant rocket systems create thrust from the chemical decomposition of chemicals ($H2O2$ or hydrazine) as they pass through a catalyst bed. They have simplified propellant handling, but have very low values of specific impulse.

Over the past sixty years, numerous chemical rocket systems have been developed and used for a wide variety of applications. Chemical rocket systems include solid propellants, cryogenic liquid propellants, storable liquid propellants, hybrid rockets, monopropellants, and cold gas rockets. The thrust on various applications have ranged from much less than a pound for attitude control to 1.5 million pounds of the F-1 engine for the Saturn V and the space shuttle main engines. However, the specific impulse is limited to several hundred lb_f-s/lb_m or less. In order to attain the high speeds required to reach the outer planetary bodies, let alone rendezvous with them, will require propulsion system efficiencies well over a 1000 lb_f-s/lb_m. These limitations make them largely inadequate for advanced space missions, particularly to the outer planets. SOA chemical propulsion systems were used as the pivot or baseline technology against which advanced propulsion technologies were evaluated during the IISTP Phase I effort.

4.3.2 Advanced Chemical

Many advanced chemical propellants are being analyzed and tested to determine their performance and applicability to in-space propulsion. The number of compounds used in the reactions typically categorizes these propellants. In addition, researchers are investigating ways to increase the Isp of current SOA chemical propellants using High Energy Density Matter (HEDM).

While the field of advanced chemical propulsion includes numerous initiatives as described below, for the purposes of the IISTP study, only O_2/CH_4 (LOX-methane) was evaluated. O_2/CH_4 is a relatively near-term technology with particular applicability to robotic planetary mission spacecraft.

4.3.2.1 Monopropellants

The most common monopropellant in use is hydrazine. It is passed through a catalyst bed, where it decomposes into nitrogen and ammonia and delivers a specific impulse of about 230 lb_f-s/lb_m. Propulsion systems of this sort are well suited to pulsed operations of short duration, such as small spacecraft attitude control.

NASA is also developing new monopropellant systems to replace the current hydrazine monopropellant systems. The monopropellants under consideration are environmentally friendly, have a higher density, and have better thermal characteristics than hydrazine. The near-term goal is to improve mission performance and greatly reduce ground operations costs. For the far-term, a very high performance (high specific impulse) system is being sought. The key to this goal is the development of a high-temperature catalyst; research in this area is underway.

4.3.2.2 Bipropellants

The bipropellant that is most often used in interplanetary spacecraft with relatively small engines is nitrogen tetroxide/monomethyl-hydrazine, commonly referred to as NTO/MMH. This combination yields an I_{sp} of 317 lb_f-s/lb_m. NASA seeks to improve performance to 326 lb_f-s/lb_m by using of a rhenium-alloyed thrust chamber, which will allow both higher operating temperatures and pressures.

NASA has also been working to improve the efficiency of LH2/LOX systems. Large pump-fed engines, like those found in the Space Shuttle main engines (SSMEs) can achieve an Isp of 450 lbf-s/lbm, while smaller pressure-fed engines can reach an Isp of 423 lbf-s/lbm. Upper stage/space engines such as the RL10IIB achieve Isp = 465 lb_f-s/lb_m. However, the high Isp may be offset by the higher structural weight, associated with the refrigeration systems required to store the cryogenic fuels for long duration missions.

Other bipropellant systems that have been investigated that use "Space storable" propellants (i.e. propellants that may be stored for extended periods in the space environment) are listed in Table 4.3.2.2-1.

Propellant	I_{sp} (lb_f-s/lb_m)
O_2/CH_4	365
ClF_5/N_2H_4	350
OF_2/C_2H_4	415
N_2F_4/N_2H_4	395
F_2/N_2H_4	415
OF_2/C_2H_6	410
OF_2/B_2H_6	420

Table 4.3.2.2-1 Isp for a variety of Bipropellant Systems

4.3.2.3 Tripropellants

There are many chemical reactions that result in a higher specific impulse than the 423 lb_f-s/lb_m that is provided by the LH2/LOX workhorse. However, many of these are unacceptable as rocket propellants because the exhaust is not a gas. Tripropellant technologies are an attempt to use these reactions by adding a third component (usually hydrogen) to the fuel and oxidizer. So far, lithium-fluorine-hydrogen and beryllium-oxygen-hydrogen mixes show the most promise for a tripropellant application.

The beryllium-oxygen-hydrogen system could generate an I_{sp} of 705 lb_f-s/lb_m and is being investigated by the U.S. Air Force. A lithium-fluorine-hydrogen system has the potential for generating an I_{sp} of 705 lb_f-s/lb_m. Early testing shows that while it has a higher combustion efficiency than the beryllium-oxygen-hydrogen system, is only allows a slight advantage over a fluorine-hydrogen bipropellant system.

4.3.2.4 High Energy Density Matter (HEDM)

In addition to the normal tripropellant approach, researchers have been looking at chemical additives that will increase the specific impulse generated by conventional bipropellant systems. These increases are achieved by adding high-energy chemicals in order to increase I_{sp}, thrust, and safety. This is not unlike adding chemicals to your car's fuel tank in order to achieve greater mileage. At the current time, HEDM is still in the basic research phase.

According to preliminary analyses that have been done at GRC, solid particles in a cryogenic carrier fluid (such as LH2) can carry HEDM additives to conventional combustion chambers. Adding these high-energy chemicals can increase the specific impulse by 19-49 $lb_f\text{-}s/lb_m$ (figured from the LH2/LOX baseline figure of 423 $lb_f\text{-}s/lb_m$). Increases in Isp are summarized in Table 4.3.2.4-1.

Carbon atoms	+ 49 $lb_f\text{-}s/lb_m$
Boron atoms	+ 31 $lb_f\text{-}s/lb_m$
Aluminum atoms	+ 27 $lb_f\text{-}s/lb_m$
Hydrogen atoms	+ 19 $lb_f\text{-}s/lb_m$

Table 4.3.2.4-1 HEDM Isp Increases

In addition to the increase to specific impulse, HEDM additives have the potential to increase propellant and vehicle density, allowing for more compact vehicles. These improvements would allow a higher percentage of deliverable payload weight to vehicle weight in future launch vehicles.

All of the chemical technologies we have discussed in this section will improve our ability to achieve orbit from the Earth's surface, but will have limited utility in traveling to other planets. The next four sections will deal with completely new technologies specifically designed for interplanetary propulsion. Some of these will be suitable for manned spacecraft, while others could be used for unmanned probes.

4.3.3 Nuclear Thermal (NTP)

The energy available from a unit mass of fissionable material is approximately 10^7 times larger than that available from the most energetic chemical reactions. A "typical" solid-core nuclear rocket engine utilizing fissionable material is shown schematically in Figure 4.3.3-1. In this engine, the propellant is heated as it passes through a heat-generating solid fuel core (nuclear reactor).

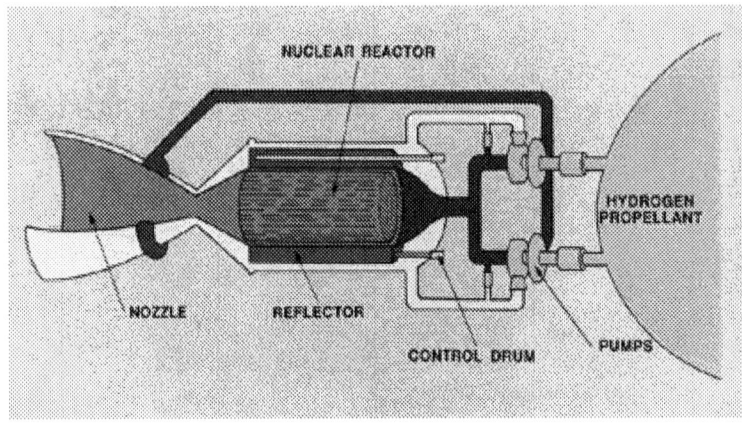

Figure 4.3.3- 1 Schematic of a "Typical" Solid-Core Nuclear rocket Engine

[10]http://siliconsky.com/sao/fit/nuclear.htm

Material constraints are a limiting factor in the performance of solid core nuclear rockets. The maximum operating temperature of the working fluid (e.g., hydrogen) must be less than the melting point of the fuel, moderator, and core structural materials. This corresponds to specific impulses of around 800 to 900 $lb_f\text{-s}/lb_m$.

4.3.4 Momentum Exchange (MX) Tether

An Earth-orbiting spinning tether system can be used to boost payloads into higher orbits. Two methods have been proposed, the (1) Spinning Boost and the (2) Swinging Boost transfer. Both rely on large orbiting tether stations with a mass 8-10X larger than the payload mass. MX tethers could provide 90% of the Earth escape velocity as well as moving satellites between LEO and GEO. However, both methods require the tether stations to be stationed in equatorial orbits and very high accuracy orbital rendezvous to be performed.

4.3.4.1 Spinning Boost (Hohmann-type) Transfer

A tether system would be anchored to a relatively large mass in LEO awaiting rendezvous with a payload delivered to orbit. The uplifted payload meets with the tether facility that then begins a slow spin-up using electrodynamic tethers (for propellantless operation) or another low thrust, high Isp thruster. At the proper moment and tether system orientation, the payload is released into a transfer orbit – potentially to geostationary transfer orbit (GTO) or Lunar Transfer Orbit (LTO). Figure 4.3.4.1-1 shows an artists rendering of the rendezvous.

Figure 4.3.4.1-1 Artists Rendering of MX Orbital Capture

The physics governing a rotating momentum exchange system is illustrated in Figure 4.3.4.1-2. Following spin-up of the tether and satellite system, the payload is released at the local vertical. The satellite is injected into a higher orbit with perigee at the release location; the orbital tether platform is injected into a lower orbit with apogee at the release location. Momentum is transferred to the satellite from the orbiting tether boost station. The satellite then enters a GTO trajectory and accomplishes the transfer in as little as 5 hours. The platform then reboosts to its operational altitude using electric thrusters. The system thus achieves transfer times comparable to a chemical upper stage with the efficiencies of electric propulsion. As shown in Figure 4.3.4.1-3, this type of system could be used to reduce launch vehicle requirements, or to increase injected payload mass, for any interplanetary mission.

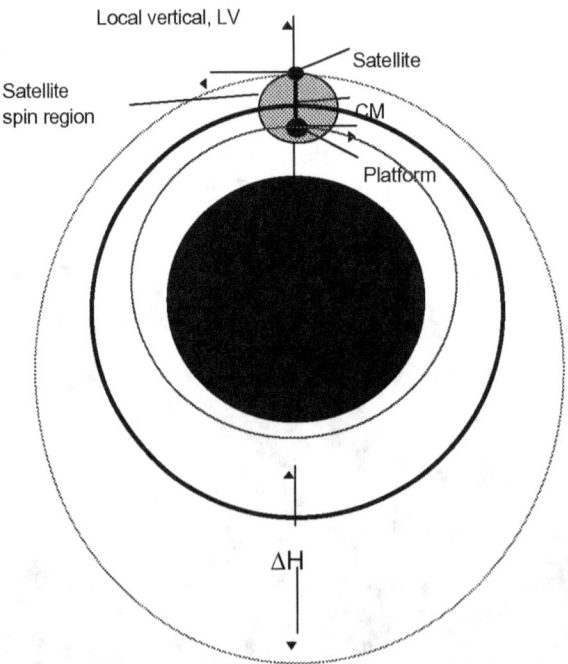

Figure 4.3.4.1-2 Orbits After Release In The "Spinning' Tether Boost Scenario

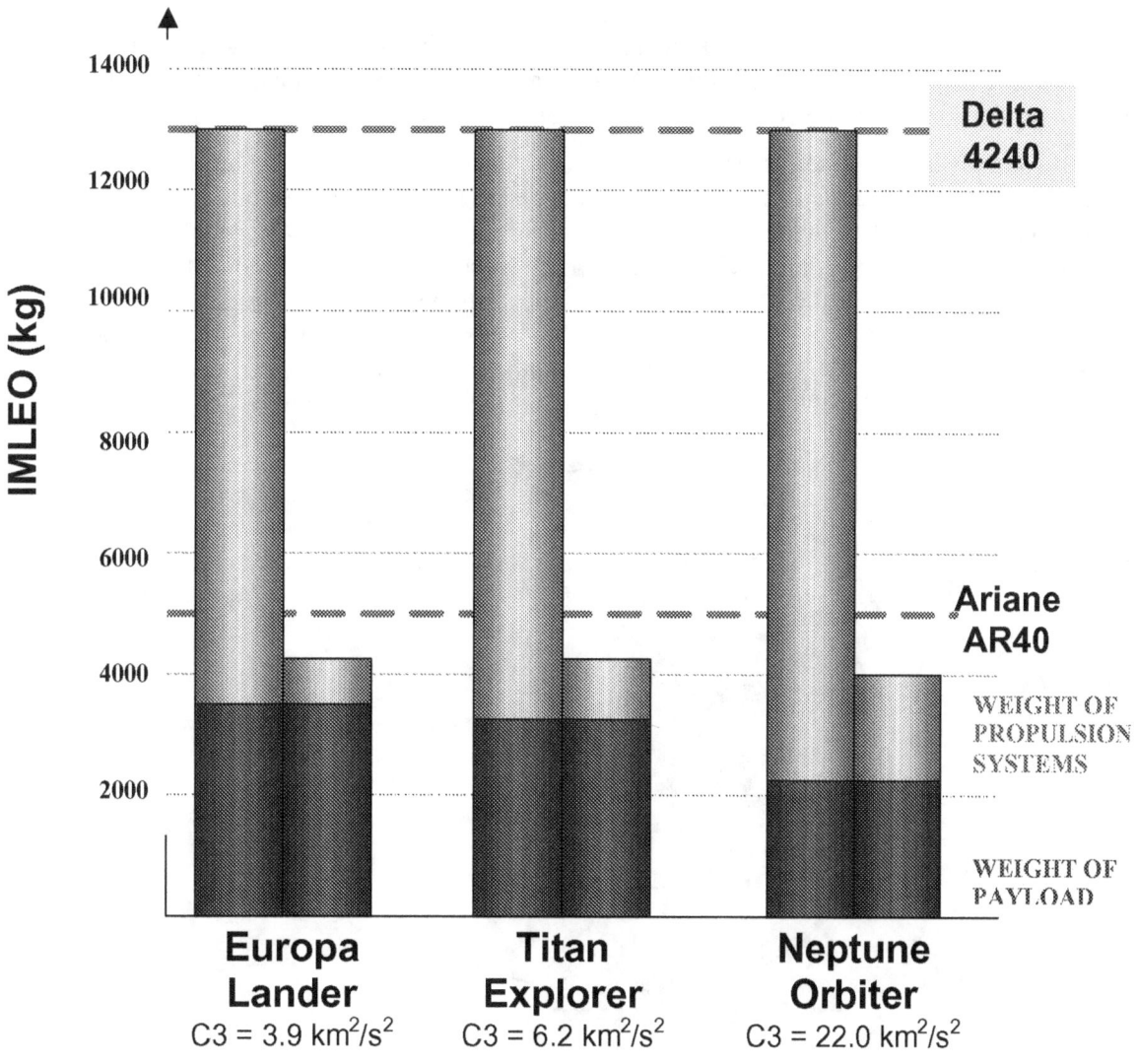

Figure 4.3.4.1-3 MX Tether Systems Reduce Launch Vehicle Size and Cost

4.3.4.2 Swinging Boost Transfer.

A long, thin, high-strength cable is deployed in orbit and set into rotation around a massive central body. If the tether facility is placed in an elliptical orbit and its rotation is timed so that the tether will be oriented vertically below the central body and swinging backwards when the facility reaches perigee, then a grapple assembly located at the tether tip can rendezvous with and acquire a payload moving in a lower orbit.

Half a rotation later, the tether can release the payload, tossing it into a higher energy orbit. This concept is termed a momentum-exchange (MX) tether because when the tether picks up and throws the payload, it transfers some of its orbital energy and momentum to the payload. The tether facility's orbit can be restored later by reboosting with propellantless electrodynamic tether propulsion or with high specific impulse electric propulsion; alternatively, the tether's orbit can be restored by using it to de-boost return traffic payloads. Figure 4.3.4.2-1 shows this method pictorially. A typical orbit for the tether platform would be 400 x 13,000 km with a tether length of 140 km.

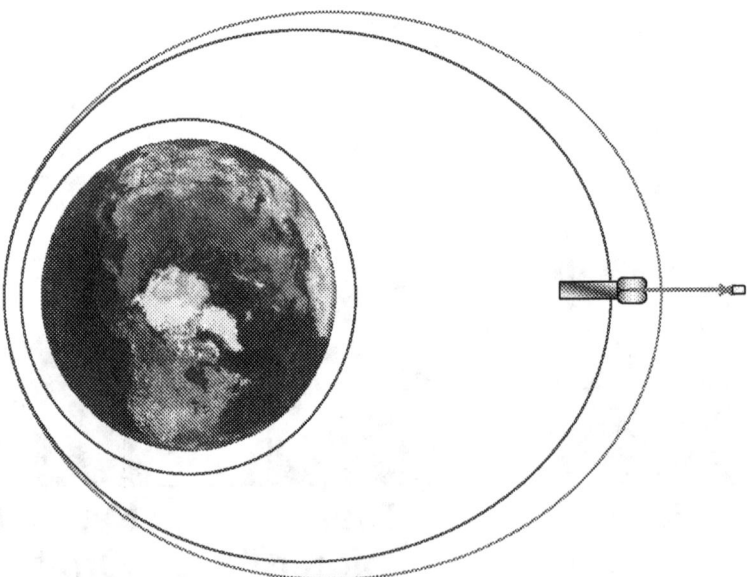

Figure 4.3.4.2-1 MX Swinging Tether at Payload Release Point

[11] http://www.tethers.com/

4.3.5 Electric Propulsion (EP)

For both chemical and electric propulsion (EP), high propellant energy results in high exhaust velocity and low mass consumption for a given thrust. In contrast to chemical rockets, on electric propulsion systems, the electric power source (solar power, nuclear power, etc.) and the thrust generating mechanisms are physically separated. On average, the energy supplied to the propellant by electricity is two orders of magnitude (~100 times) higher than the energy supplied in chemical propulsion through a chemical reaction. However, electric propulsion is power limited by the rate of energy conversion (e.g., solar or nuclear energy into electric energy).

Spacecraft using electric propulsion systems for space missions require less propellant at launch and on orbit than chemical systems, thereby reducing launch costs while increasing the payload of the launch vehicle and spacecraft, and by providing mission engineers with greater design flexibility. Electric propulsion devices are capable of generating low thrust for long periods of time. The final velocities are at least the same or higher than that achieved with chemical propulsion, because electric rockets accelerate much longer. For planetary missions, significant time savings can be achieved with electric thrusters since time-consuming (long travel times, timing for a particular rendezvous launch window, etc.) gravity assist maneuvers to reach high final velocities are not required. In contrast to chemical propulsion, small quantities of propellant mass are expelled through the thruster at extremely high velocities. Figure 4.3.5-1 shows two artists renderings of EP systems.

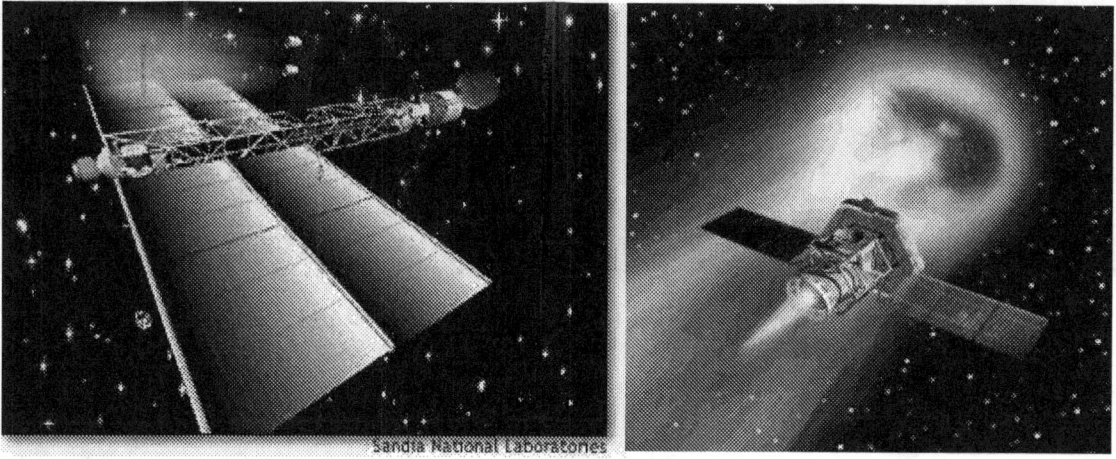

Figure 4.3.5-1 Electric Propulsion Concepts

IISTP Phase I Final Report
September 14, 2001

NASA is pursuing technologies to increase the performance of electrostatic thrusters by going to higher power levels and by increasing the Isp on a system level. Figure 4.3.5-2 illustrates the mission benefit of using electric propulsion to increase the payload mass fraction.

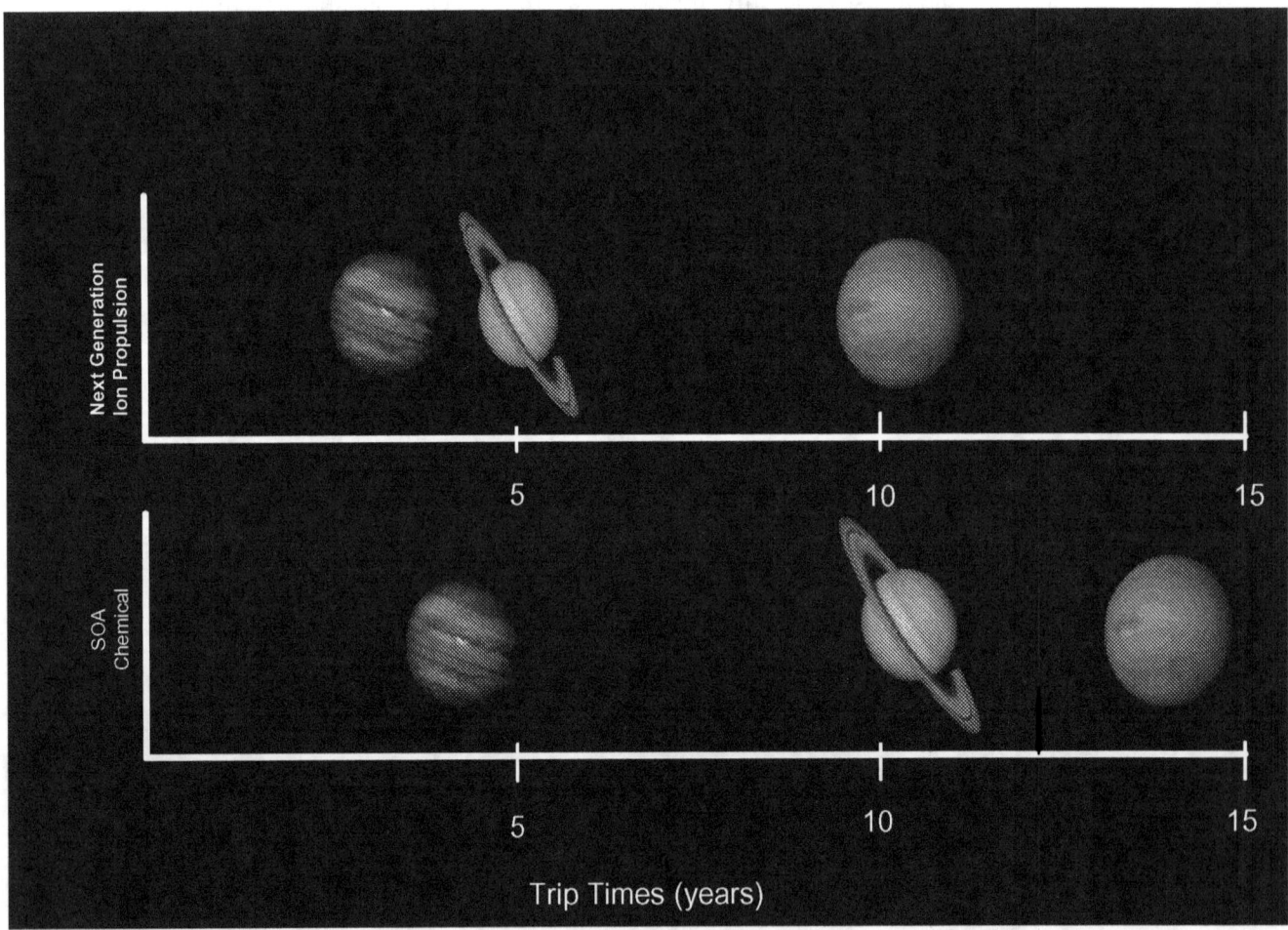

Figure 4.3.5-2 EP Systems Can Significantly Reduce Trip Times

Electric propulsion is most broadly defined as the acceleration of propellants by electrical heating, electric body forces, and/or magnetic body forces. This leads to a natural division of the three forms of electric propulsion:

1) Electrothermal thrusters- the propellant gas is electrically heated and thermodynamically expanded through a nozzle. Common examples include resistojets and arcjets.

2) Electrostatic thrusters- the propellant is ionized and the resulting ions are accelerated through an electric potential. Common examples include Hall effect and Ion type thrusters.

3) Electromagnetic thrusters- utilize electric and magnetic body forces to accelerate ions. The Variable Specific Impulse Magnetoplasma Rocket (VASIMR) thruster falls into this class of EP thrusters.

For the class of missions considered during the IISTP Phase I effort, only the Electrostatic and Electromagnetic electric propulsion systems were considered viable ISP technology candidates.

4.3.5.1 Electrothermal Systems

Electrothermal thrusters electrically add energy to a suitable propellant and expand the hot gases through a supersonic nozzle, thus converting electrical energy into kinetic energy. Thrust is generated by thermally expanding the hot propellant in a converging-diverging nozzle. Thrust and specific impulse are limited by the thruster material properties.

4.3.5.2 Electrostatic Systems

In contrast to electrothermal propulsion devices, electrostatic thrusters exert electric body forces on charged particles via electrostatic fields. The direct electric acceleration of charged particles eliminates thermal limitations inherent in solid wall material properties, thus lifting restrictions on thrust and specific impulse. The characterization of electrostatic systems is based on the production mechanisms of charged particles; these can be summarized as electron bombardment thrusters, radio-frequency ion thrusters, ion contact thrusters, and field emission thrusters. The principle of operation is illustrated in Figure 4.3.5.2-1.

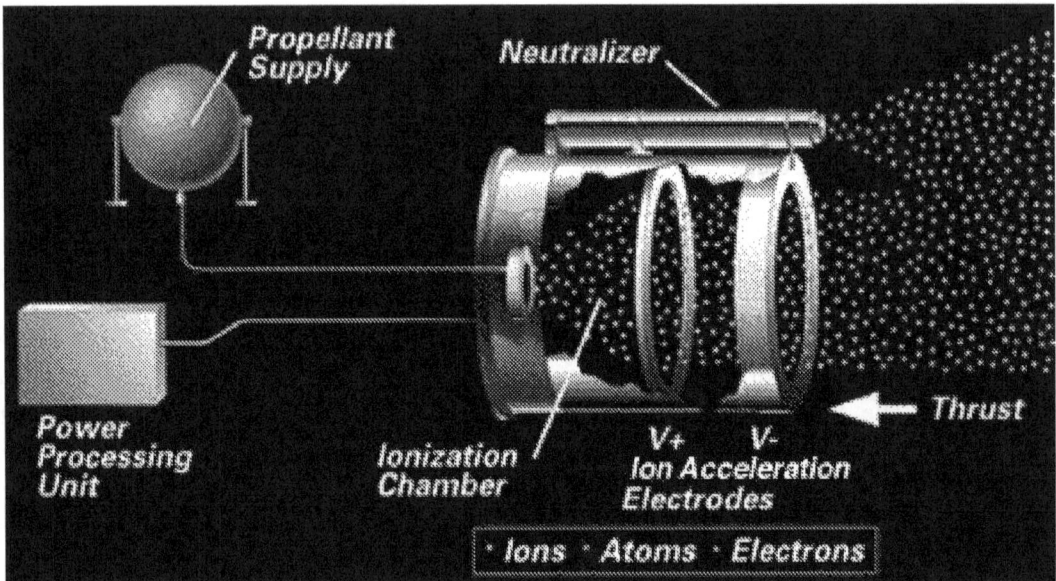

Figure 4.3.5.2-1 Basic Electrostatic Operation

Ions are created using one of the above mechanisms and accelerated in the electric field between the positive ion source and a negative grid electrode. At the exit plane of the thruster, a neutralizer supplies electrons to the ion stream producing a beam of zero net charge whose exit velocity is governed by both the potential difference between neutralizer and ion source, and the mass-to-charge ratio of the ions. The ion mass-to-charge ratio is very important since the thrust per unit area increases with the square of the mass-to-charge ratio.

4.3.5.2.1 Ion Thrusters

Ion thrusters achieve very high specific impulse by accelerating charged particles across a potential difference using electrostatic force fields. Ion propulsion is being used by commercial telecommunication satellites and has been demonstrated as a primary spacecraft propulsion system by the NASA Solar Electric Technology Application Readiness (NSTAR) demonstration on the Deep Space 1 (DS1) mission. Under the circumstances for which grid Ion propulsion is appropriate, a spacecraft can reach a final velocity of approximately ten times greater than that of a spacecraft using chemical propulsion. Because the Ion propulsion system, although highly efficient, is very gentle in its thrust, it cannot be used for any application in which a rapid acceleration is required. With patience, the Ion propulsion system on DS1 imparts about 3.6 km/s to the spacecraft. To undertake the same mission with a chemical propulsion system would require a more expensive launch vehicle and a larger spacecraft to accommodate a large tank for the chemical propellants.

The electrical energy to power these devices can be provided by a solar power source, such as solar photovoltaic arrays, or a nuclear power source, such as a space based nuclear reactor. Nuclear power can be accomplished using a fission reactor or through the use of radio-isotopes batteries. In either case, the mass of the EP power source partially offsets the propellant mass savings from the high specific impulse, thus a highly efficient, low mass power source is essential for the successful implementation of any electric propulsion technology. Nuclear Electric Propulsion (NEP) systems are of great interest in those missions to the outer solar system, where solar power is no longer efficient and Solar Electric Propulsion (SEP) systems are not feasible. SEP and NEP Ion Systems with various power ratings were analyzed during Phase I. Figure 4.3.5.2.1-1 shows a schematic of a typical Ion thruster.

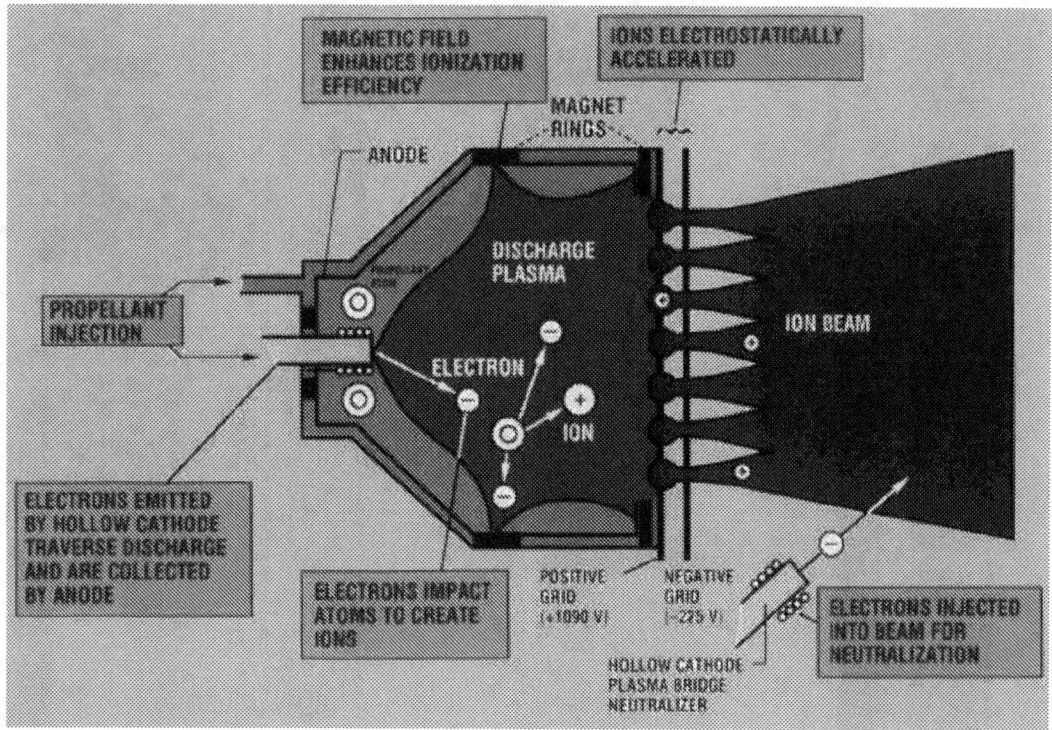

Figure 4.3.5.2.1-1 Typical Ion Thruster

4.3.5.2.2 Hall Thrusters

Hall thrusters use an axial electric field to accelerate ions, similar to Ion thrusters. Combining a radial magnetic field with this generates an azimuthal Hall current. This current interacts with the radial magnetic field producing a volumetric (j x B) accelerating force on the plasma. As with grid ion thrusters, Hall thrusters can be categorized according to their respective power sources (i.e. solar or nuclear). Solar Electric Propulsion (SEP) Hall Systems and Nuclear Electric (NEP) Hall Systems with various power ratings were analyzed. Figure 4.3.5.2.2-1 shows a simplified schematic of a Hall Thruster.

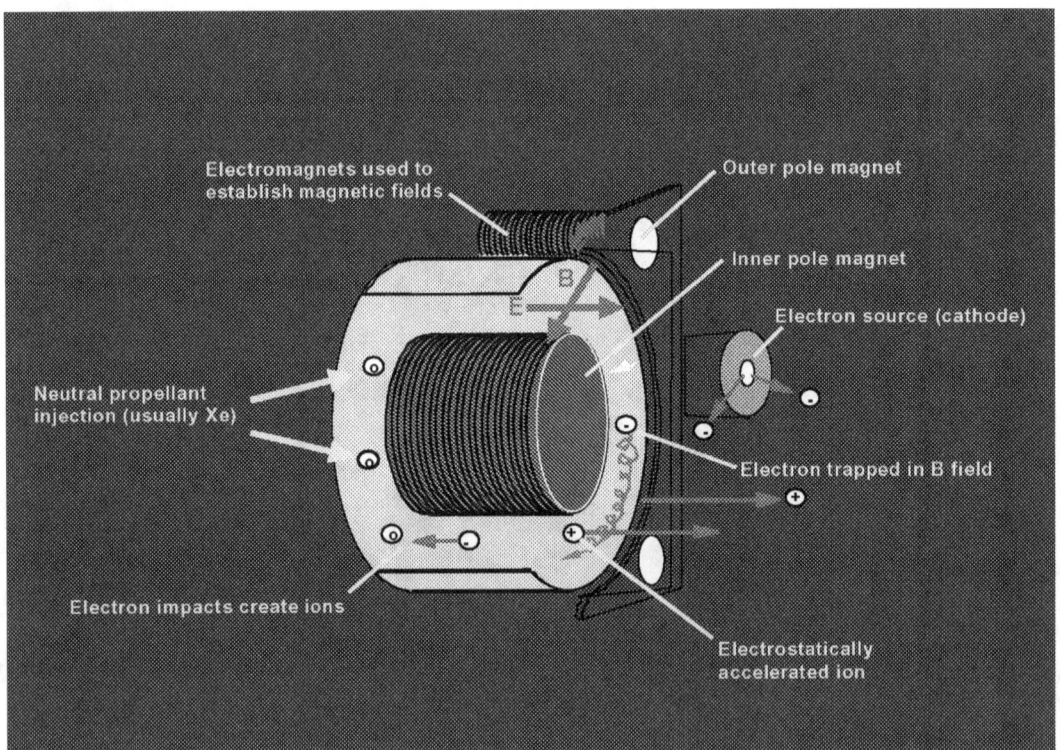

Figure 4.3.5.2.2-1 Schematic of a Hall Thruster

4.3.5.3 Electromagnetic Propulsion

The electromagnetic propulsion system evaluated for this effort was the Variable Specific Impulse Magnetoplasma Rocket (VASIMR). The VASIMR system consists of three major magnetic cells, denoted as "forward," "central," and "aft." This particular configuration of electromagnets is called an asymmetric mirror. The forward end-cell involves the main injection of gas to be turned into plasma and the ionization subsystem; the central-cell acts as an amplifier and serves to further heat the plasma. The aft end-cell ensures that the plasma will efficiently detach from the magnetic field. Without the aft end-cell, the plasma would tend to follow the magnetic field and provide only a small amount of thrust. With this configuration, the plasma can be guided and controlled over a wide range of plasma temperatures and densities.

To operate the VASIMR, neutral gas (typically hydrogen) is injected at the forward end-cell and ionized. Then it is heated to the desired temperature and density in the central-cell, by the action of electromagnetic waves, similar to what happens in microwave ovens. After heating, the plasma enters a two-stage hybrid nozzle at the aft end-cell where it is exhausted to provide modulated thrust. Figure 4.3.5.3-1 provides a schematic of the VASIMR propulsion system.

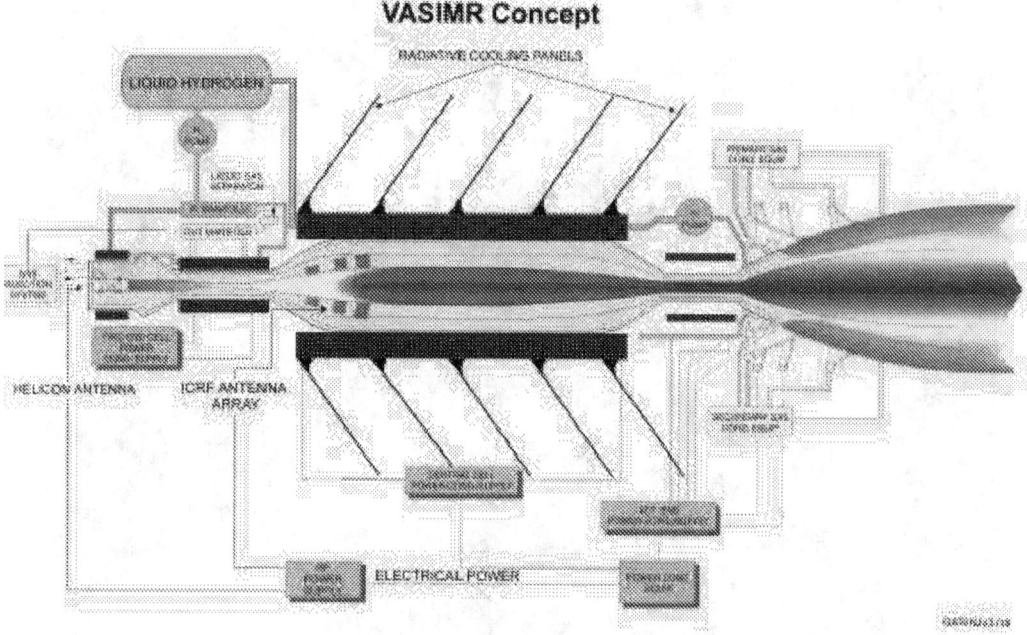

Figure 4.3.5.3-1 VASIMR Propulsion System

[12] http://spacsun.rice.edu/aspl/vasimr.htm

4.3.6 Solar Sails

A solar sail is a propulsion concept that makes use of a flat surface of very thin reflective material supported by a lightweight deployable structure. Solar sails accelerate under the pressure from solar radiation (essentially a momentum transfer from reflected solar photons), thus requiring no propellant. Since a solar sail uses no propellant, it has an effectively infinite specific impulse; however, the thrust-to-weight ratio is very low, typically between 10^{-4} to 10^{-5} (for the 9 N/km^2 solar pressure at Earth's distance from the Sun).

In the near-term, deployable sails will be fabricated from materials such as Mylar or Kapton coated with about 500 Angstroms of aluminum. The thinnest available Kapton films are 7.6 microns in thickness and have an areal density of approximately 11 g/m^2. Sails thinner than this, made from conventional materials, have the potential to rip or tear in the deployment process. Recent breakthroughs in composite materials and carbon-fiber structures may make sails of areal density less than 1 g/m^2 a possibility. The reduced sail mass achieved this way may allow much greater acceleration, greater payload carrying capability, and reduced trip time. Figure 4.3.6-1 shows a conceptual solar sail being used for primary propulsion.

Figure 4.3.6-1 Solar Sail Concept

[13] http://www.howstuffworks.com/solarsail.htm

Solar sails can substantially reduce overall trip time and launch mass for many types of missions. Solar Sails are an ideal application for earth/sun station keeping satellites. Another example, the proposed Interstellar Probe (ISP) mission, cannot be practically achieved without solar sails or nuclear electric propulsion. The reduction of trip times possible with a solar sail is illustrated in Figure 4.3.6-2. The baseline solar sail concept developed to meet ISP requirements assumes a spin-stabilized sail with an areal density of 1 g/m2 (including film and structure), and a diameter of approximately 400 meters with an 11-meter wide central opening. The spacecraft module would be centered in the central aperture of the sail. The total spacecraft module mass supported in the sail would be approximately 180 kg. The ISP sail craft would be used on a heliocentric trajectory from Earth escape inbound to a 0.25 AU perihelion, then outbound to 5 AU, where the sail would be jettisoned to minimize interference with acquisition of scientific data and communication. A single Delta II class launch vehicle would be used to deliver the sail-craft to an Earth-escape trajectory.

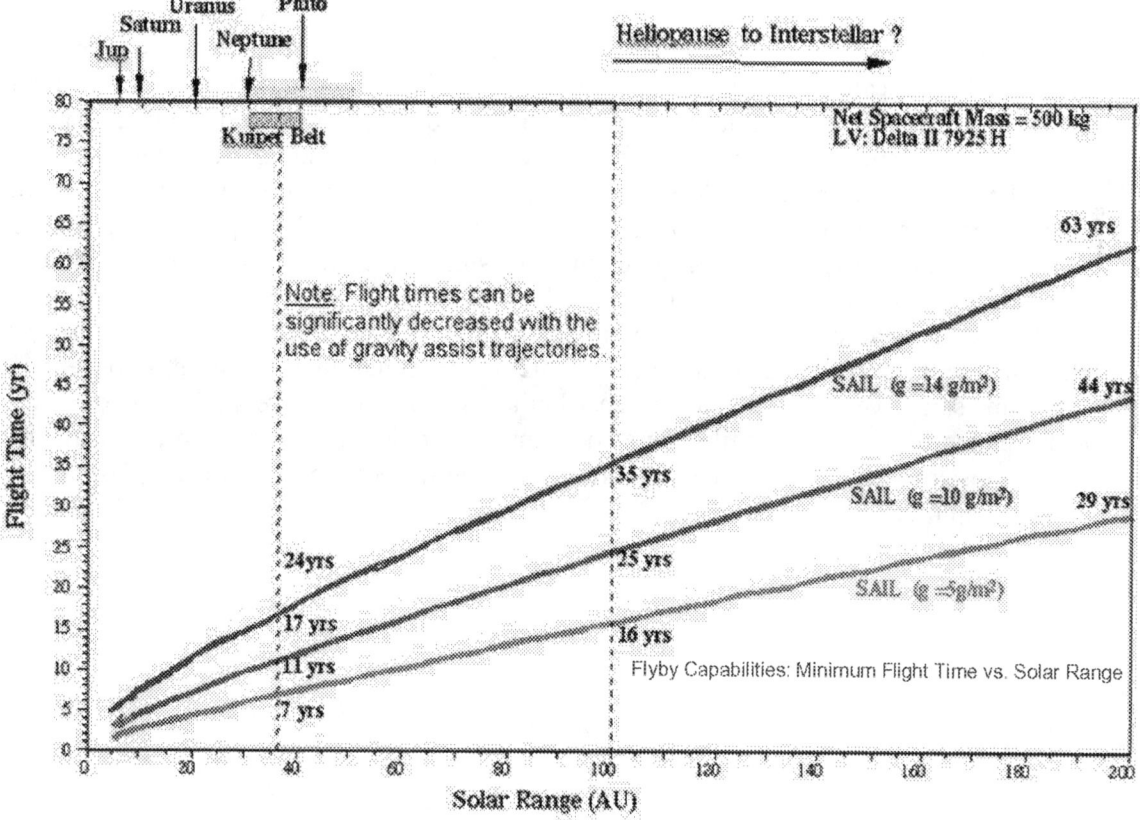

Figure 4.3.6-2 Solar Sail Dramatically Reduces Trip Times

4.3.7 Plasma Sails (M2P2)

A novel new approach to spacecraft propulsion using a virtual sail composed of low energy plasma might harness the energy of the solar wind to propel a spacecraft anywhere in the solar system and beyond. Such plasma sails will effect their momentum transfer with the plentiful solar wind streaming from the sun, requiring very little propellant. Plasma sails use a plasma chamber attached to a spacecraft as the primary propulsion system. Solar cells and solenoid coils would power the creation of a dense magnetized plasma, or ionized gas, that would inflate an electromagnetic field up to 19 kilometers in radius around the spacecraft. (In the future, fission power could be used.) The field would interact with and be dragged by the solar wind. Creating this virtual sail will be analogous to raising a giant physical sail and harnessing the solar wind, which moves at 780,000 to 1.8 million miles an hour.

Tests of the plasma sail concept are ongoing at the Marshall Space Flight Center (MSFC) and The University of Washington. Figure 4.3.7-1 shows trapped plasma on closed field lines extending 2 meters into a large vacuum chamber during a recent series of tests at MSFC. Specifically, the image shows inflation of the helium gas feeding the helicon plasma source. The luminosity results from the ions colliding with and exciting residual gas in the chamber. Visible closed field lines were seen extending 2-3 meters into the chamber from a 20 cm coil. There is evidence, such as in this image, that much more distant field lines were closed. In particular, the arc seen extending downward from the coil would not be expected to be there unless plasma ejected from the top followed high latitude field lines to the "southern" hemisphere of the coil. That puts closed field lines perhaps entirely across the chamber or about 5 meters. Thrust tests, using a Hall Thruster to simulate the solar wind, are planned in 2002 – 2003 timeframe. An artist's concept of a plasma sail driven spacecraft flying past Jupiter is shown in Figure 4.3.7-2. Depending on the size of the plasma sail generated, significant reductions in trip times for all in-space missions can be achieved. Figure 4.3.7-3 shows how the size of the plasma sail effects trip times.

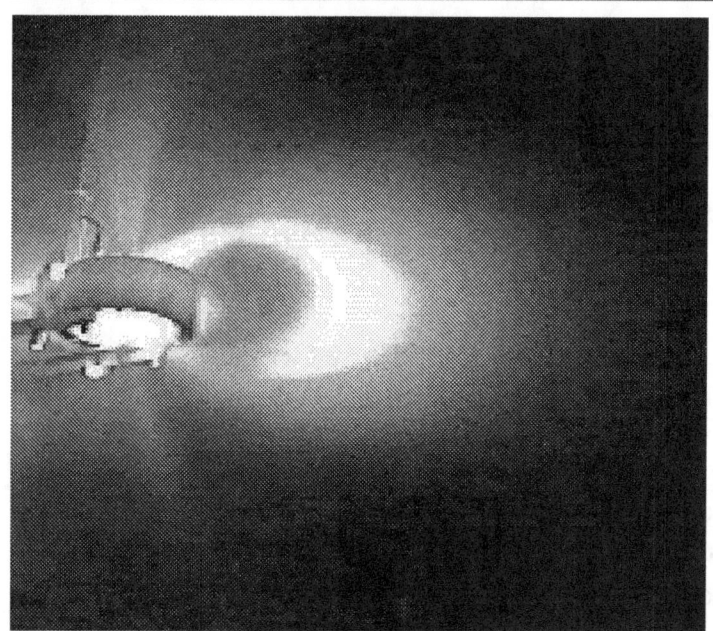

Figure 4.3.7-1 Plasma Sail Inflation Demonstration in 2000

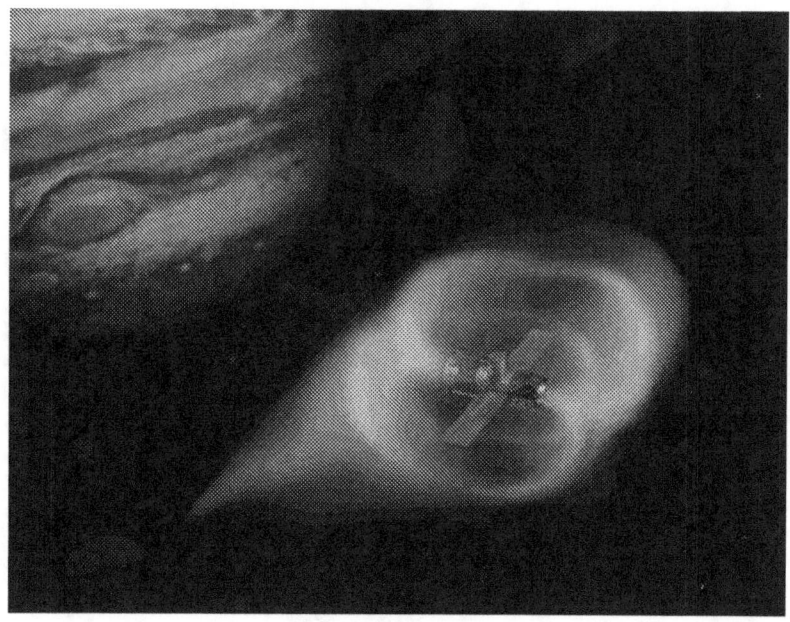

Figure 4.3.7-2 Spacecraft Using a Plasma Sail

[14] http://spike.geophys.washington.edu/Space/SpaceModel/M2P2/

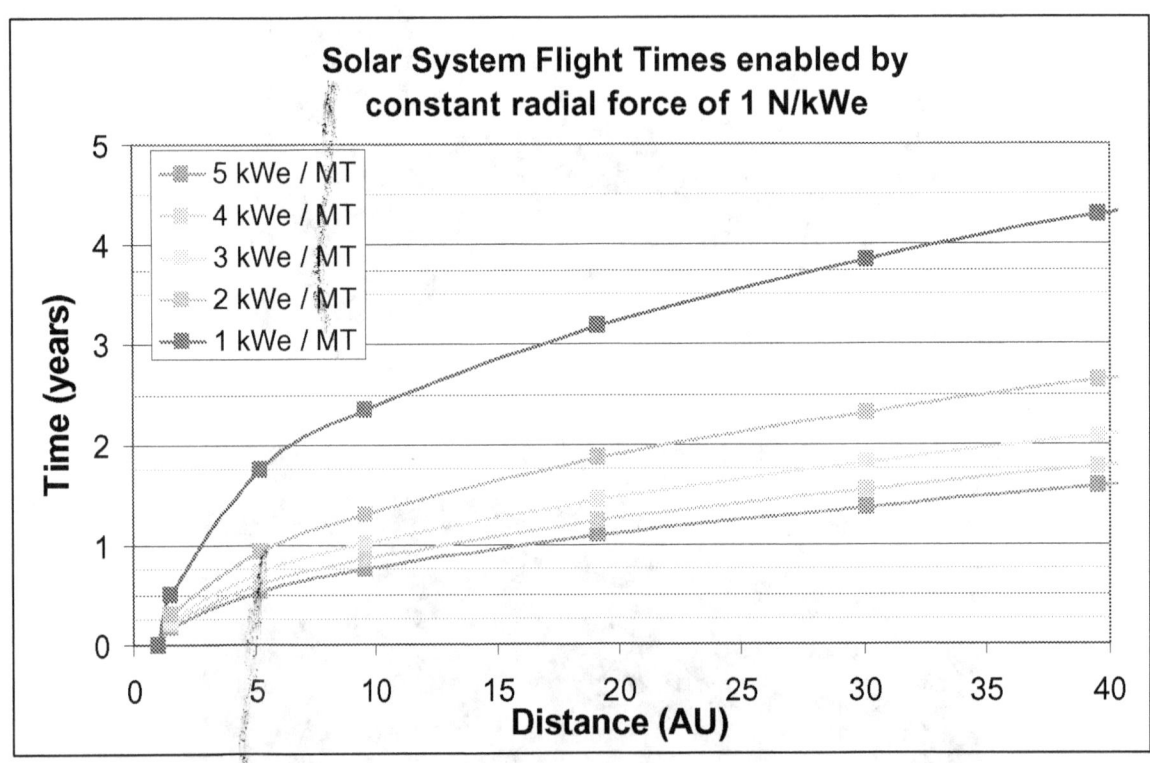

Figure 4.3.7-3 M2P2 Sail Trip Time Reductions

4.3.8 Solar Thermal

Solar thermal propulsion (STP) effectively bridges the performance gap between chemical and electric propulsion by offering higher Isp's (= 800 - 1000 secs) than chemical options (= 300 - 500 secs) and higher thrust-to-weight ratios than electric systems. STP requires only one propellant and combines medium thrust with moderate propellant efficiency to enable relatively short 30-day trips from low Earth orbit to geostationary Earth orbit.

The propulsion system of a solar thermal-powered spacecraft consists of three basic elements: a Concentrator which focuses and directs incident solar radiation, a Thruster/absorber which receives solar energy, heats and expands propellant (hydrogen) to produce thrust, and a Propellant system which stores cryogenic propellant for extended periods and passively feeds it to the thruster/absorber. Figure 4.3.8-1 provides a simplified description of the operation of a solar thermal propulsion system.

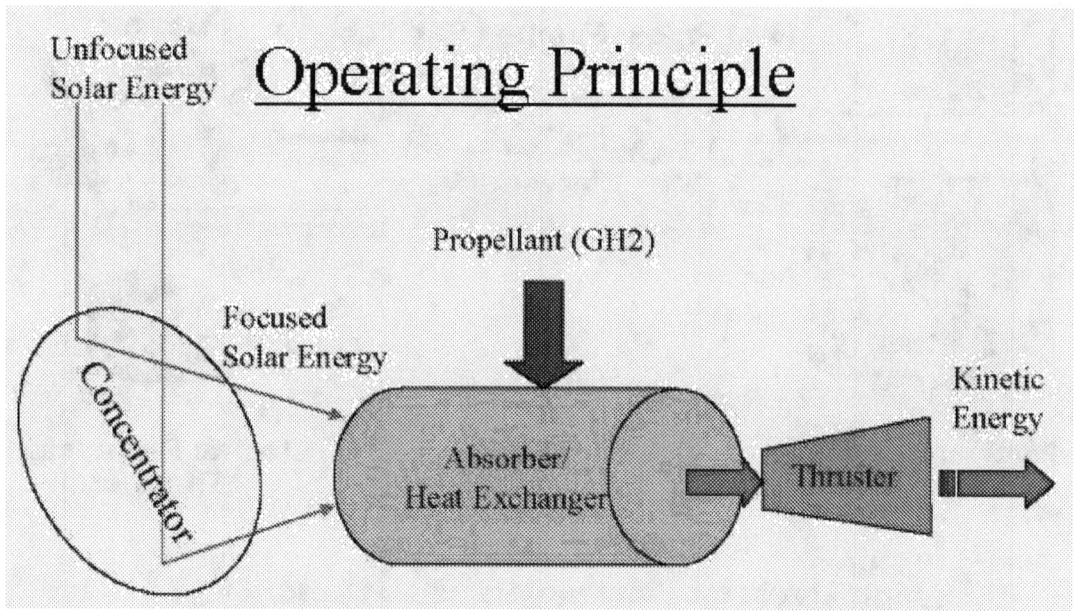

Figure 4.3.8-1 Solar Thermal Propulsion Operating Principle

[15] http://www.msfc.nasa.gov/STD/propulsion/research/solar/

4.3.9 Nuclear Thermal Propulsion/Nuclear Electric Propulsion Bimodal

The Nuclear Thermal Propulsion/Nuclear Electric Propulsion (NTP/NEP) bimodal system uses the NTP engine for maneuvers in a high-gravity field, where its high thrust-to-weight ratio minimizes gravity losses and trip time. Once outside of a planet's gravity well, the system uses the nuclear reactor to produce electricity for a NEP engine that is well suited for interplanetary transfers, due to its low T/W ratio and high I_{sp}.

The mission benefits of this approach are highly mission dependent, because there is a trade-off between the high T/W (e.g., vehicle T/W>0.1) and relatively low I_{sp} (e.g., 800-1000 lb_f-s/lb_m) of the NTP mode, and the low T/W (e.g., vehicle T/W<10^{-3}) and relatively high I_{sp} (e.g., 2000-5000 lb_f-s/lb_m) of the NEP mode. Figure 4.3.9-1 shows a simple schematic of a NTP/NEP bimodal propulsion system.

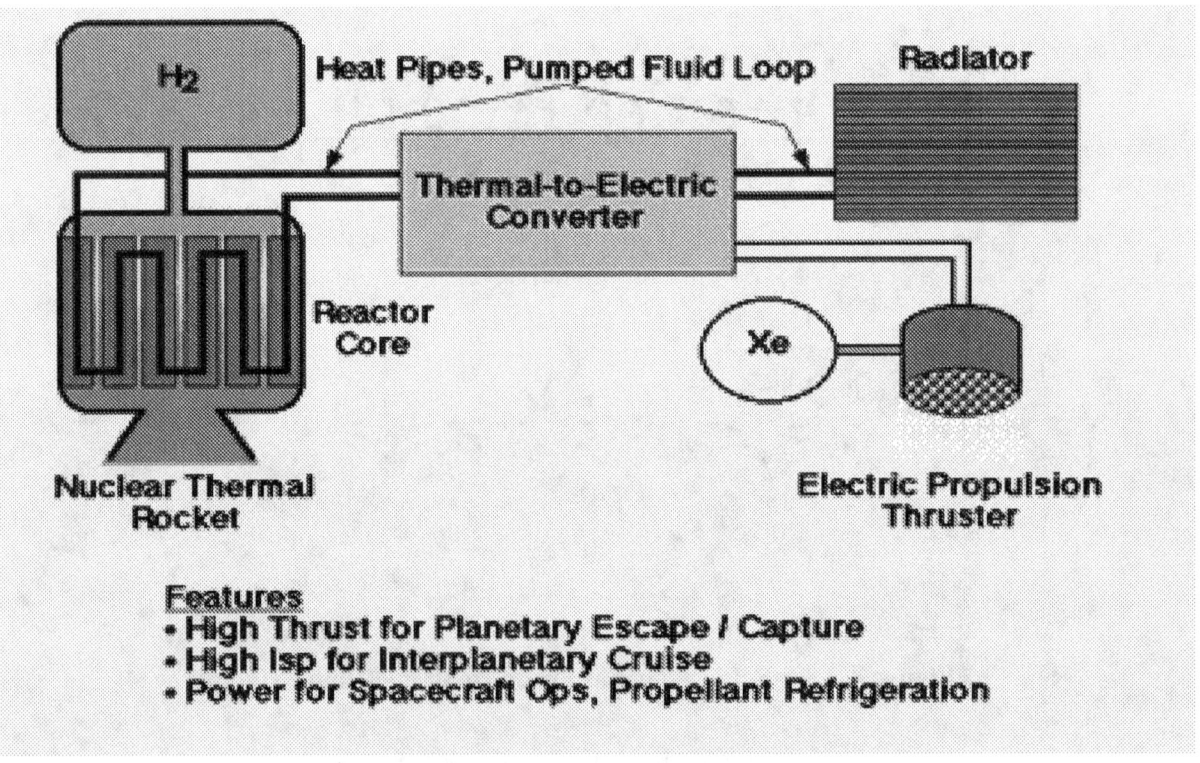

Figure 4.3.9-1 NTP/NEP Bimodal Operating Principle

[16] http://siliconsky.com/sao/fit/electric.htm

4.3.10 Aerocapture

Aerocapture relies on the exchange of momentum with a planetary atmosphere to achieve thrust, in this case a decelerating thrust leading to orbit capture. Aerocapture has not yet been demonstrated, though it is very similar to the flight-proven technique of aerobraking, with the distinction that aerocapture is employed to reduce the velocity of a spacecraft flying by a planet so as to place the spacecraft into orbit about the planet. This technique is very attractive for planetary orbiters since it permits spacecraft to be launched from Earth at high speed, providing a short trip time, and then reduce the speed by aerodynamic drag at the target planet. Without aerocapture, a large propulsion system would be needed on the spacecraft to perform the same reduction of velocity. Possible impacts would include reductions in the delivered payload mass, increases in the size of the launch vehicle (to carry the additional fuel required for planetary capture) or simply making the mission impossible due to the tremendous propulsion requirements. Figure 4.3.10-1 shows various conceptual aerocapture techniques.

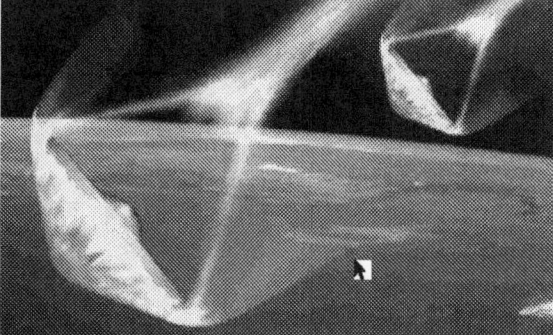
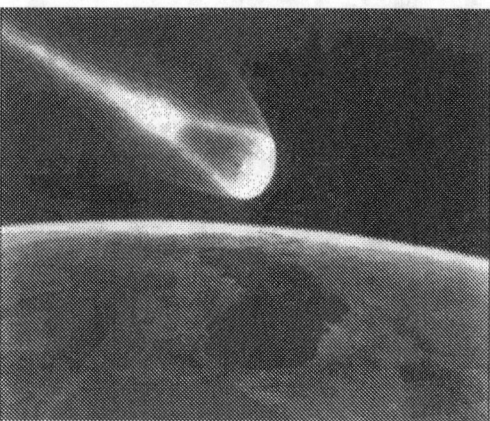

Figure 4.3.10-1 Various Aerocapture Techniques

The aerocapture maneuver begins with a shallow approach angle to the planet, followed by a descent to relatively dense layers of the atmosphere. Once most of the needed deceleration is reached, the vehicle maneuvers to exit the atmosphere. To account for the inaccuracies of the atmospheric entering conditions and for the atmospheric uncertainties, the vehicle needs to have guidance and control as well as maneuvering capabilities. Most of the maneuvering is done using the lift vector that the vehicle's aerodynamic shape (i.e., lift-to-drag ratio, L/D) provides. Upon exit, the heatshield is jettisoned to minimize thermal problems and a short propellant burn is required to raise the orbit periapsis. Given the communication time delay resulting from the mission distances from Earth, the entire operation requires the vehicle to operate autonomously while in the planet's atmosphere. Figure 4.3.10-2 shows the propulsion system mass savings that are possible with an aerocapture system.

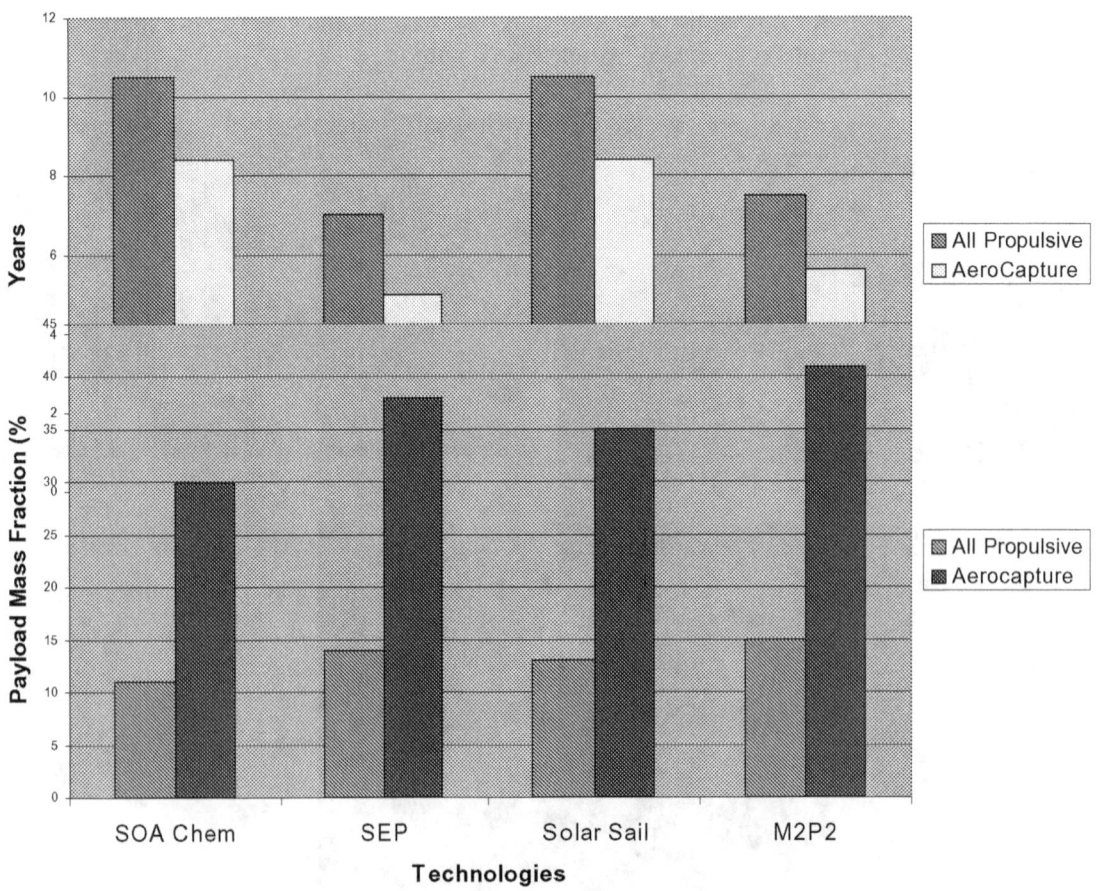

Figure 4.3.10-2 Aerocapture Reduces Propulsive Requirements for Capture Maneuvers

4.4 Step Four: Figures of Merit (FOM) Weighting

The FOMs were tailored for each mission category through the application of weights. Some FOMs were extremely important for some mission categories and not even applicable for others. Therefore, for each mission category the FOMs were weighted by the appropriate Enterprise on a scale of "0 to 10", where "0" indicated "not applicable" and "10" indicated "of highest importance".

The "0 to 10" weighting scale was adopted from the highly successful Kepner-Tregoe[17] (K-T) method for decision making used throughout government and industry over the past forty years. Once all "not applicable" criteria are weighted "0", the K-T method suggests that within each of the five FOM categories the most important criteria be identified first and be weighted a "10". Next, the remaining criteria should be weighted in relative importance to the most important criteria on a scale of 1 to 10 within the FOM category. It is important to note, the criteria are not ranked, rather, a "pair-wise" comparison of each criterion to those criteria weighted a "10" are made. Given these guidelines, each FOM was weighted by the appropriate Enterprise, and the resulting weights for the mission categories analyzed are shown in Table 4.4-1.

Unfortunately, the K-T method was not well understood by some of those assigning the weights and some of the K-T guidelines were not strictly adhered to. For example, a weight of "10" was not assigned to any of the FOMs in the "Technical", "Reliability/Safety", Cost" or "Schedule" FOM categories for the "Earth Vicinity" mission category. This did not affect the end result since all scores were normalized to 100 within each FOM category (see Section 4.5).

The FOM weights were not disclosed by the MRT to any other members on IISTP team (with the exception of a select few IAG members) until the TT, ST and CT completed all scoring. This ensured that members of the scoring teams could score each FOM independently without regard to their relative importance to one another.

IISTP Phase I Final Report
September 14, 2001

FOM CATEGORY		FIGURES OF MERIT	Earth Vicinity	Solar System & Beyond	HEDS-Mars
PERFORMANCE	1	Payload Mass Fraction	10	9	5
	2	Trip Time	5	10	10
	4	Time on Station	5	0	0
	19	Propulsion System Launch Mass & Volume	0	10	1
TECHNICAL	3	Operational Complexity	5	9	5
	5	Propellant Storage Time	5	9	5
	6	Station Keeping Precision	2	0	0
	14	Crew Productivity	0	0	5
	15	Sensitivity to Malfunctions	8	10	5
	16	Sensitivity to Performance Deficiencies	7	10	5
	17	Enable In-Space Abort Scenarios	2	0	7
	18	Crew Exposure to In-Space Environments	5	0	3
RELIABILITY/ SAFETY	31	Pre-Launch Environmental Hazards & Protection	5	10	0
	41	In-Space Environmental Hazards & Protection	8	10	2
	42	Crew Exposure & Safety	0	0	8
	43	Payload Exposure & Protection	8	8	8
	51	Relative Reliability Assessment	8	10	10
	53	Operating Life	7	10	0
COST	61	Technology Advancement Cost	8	9	2
	62	Mission Non-Recurring Cost	8	9	10
	67	Operational Cost	9	10	7
	68	Mission Recurring Cost	9	10	7
SCHEDULE	81	Total Development Time	5	10	10
	82	Special Facility Requirements	5	9	3
	83	Architectural Fragility	5	9	5
	84	Maturity (TRL Level)	5	8	10

Table 4.4-1 IISTP Phase I FOM Weights

4.5 Step Five: Evaluation and Scoring of Candidate ISP Technologies

This section describes in detail the process of evaluation and scoring of the candidate ISP technologies, in order to provide a record of what was done, how it was done and in some instances why. Detailed technical information from the mission analyses is presented in Appendix D. Selection of the scoring methodology is described, as is the scoring process itself. Finally, compilation and presentation of the results in a form useful to decision makers is described.

4.5.1 Mission Analyses

Time, resource and technology informational constraints severely limited the depth and extent of the mission analyses performed during IISTP Phase I. In general, mission analyses included performing trajectory analysis, applying propulsion performance and sizing algorithms, and determining initial launch conditions for the mission. For low-thrust systems, trajectory analyses were parametric in terms of specific impulse, and in some cases mass-to-power ratio performance. (Low-thrust trajectories vary depending on these parameters, and feasible and/or reasonable payload mass fraction and trip times must be determined jointly with Isp, mass-to-power ratio, and trajectory.)

The Neptune Orbiter mission was the pathfinder for all scoring, and nearly every ISP system was evaluated for that mission.

4.5.2 Scoring Methodologies

Once mission analyses were completed, the scoring teams were provided with guidelines in the FOM Dictionary for scoring each of the candidate ISP technologies. Scoring guidelines were based on a non-linear scale of 0, 1, 3, or 9 representing none, weak, moderate or strong satisfaction of the FOM, respectively. This scheme was adopted from similar applications using Quality Function Deployment[18,19,20] (QFD), widely used throughout the world since 1966. Other scoring methodologies such as Analytic Hierarchy Process[21,22,23] (AHP) were considered but did not seem appropriate given the nature of the Phase I process. Specifically, on each mission AHP would require the scoring teams to perform pair-wise comparisons of every candidate technology against a baseline or pivot technology. AHP redundancy does promote consistent scoring. However, the number of judgments required to perform redundant pairwise comparisons can be very large if there are a large number of FOMs and/or a large number of candidate technologies.

The number of pair-wise judgments required when using AHP is given by:

$$AHP = m \frac{n!}{2(n-2)!}$$

Where

n = number of ISP systems

and

m = number of FOM

For example, to evaluate the Neptune Orbiter mission,

n = 20 and m = 26

Therefore, AHP would require nearly 5,000 judgments be made.

The number of judgments required when using QFD is given by:

$$QFD = n \times m$$

For the Neptune Orbiter evaluation, QFD allows the same evaluations to be made with a little over 500 judgments. This is nearly an order of magnitude difference in the overall number of judgments required for the mission. Sensitivity analyses indicated the consistency achieved using QFD equaled that for AHP for these analyses.

4.5.3 Scoring Activities

The ST and TT worked together since there was overlapping membership, and the considerations applied to derive scores for several of the FOM had common factors. The CT for the most part worked independently, but reviewed its findings with the ST and TT to ensure reasonable consensus.

Teleconferences were scheduled Tuesdays and Thursdays by the ST Lead and usually lasted about two hours. Applicable performance data, strawman scoring when available, and other information were distributed in advance by e-mail. Scoring was accomplished according to the guidelines given in the FOM Dictionary.

4.5.3.1 Strawman scoring

Strawman scoring was used at the beginning of the scoring process (using the mission analysis results for the Neptune Orbiter mission) to test and refine the process. Strawman scoring was also used throughout the process to facilitate scoring in the telecon by providing a "scoring starting point".

4.5.3.2 Data Presentation

Trajectory, performance and other relevant data, such as available briefings and reports, were developed in advance of each scoring telecon. The preparer or other knowledgeable person would review the data prior to beginning each scoring session.

4.5.3.3 Advocacy and/or Expert Input

At least one expert on each propulsion system was on hand for the scoring discussions, to (1) ensure correct interpretation of the technology capabilities, (2) answer questions, particularly as to how well the technology satisfied each FOM, and (3) to generally act as an advocate for the technology. The role of the ST members was to act in a neutral evaluation role, and to raise issues that might affect each technology's score.

4.5.3.4 Reaching Consensus

A goal during the scoring telecons was to achieve consensus whenever possible. At times, lengthy discussions of particular merits of a technology ensued. Normally, more than one technical expert was knowledgeable on the technology being considered, and multiple opinions would be offered. If a consensus could not be reached, the ST Lead would make the scoring decision based on a majority view, and add a note to that effect including the subject of the dispute to the score. Out of almost 2500 scores assigned, less than ten were disputed. A sensitivity analysis, later applied, determined that none of the disputed scores had an effect on the overall final prioritization of ISP technologies.

4.5.3.5 Recording of Results

The ST Lead recorded all scores in a spreadsheet format that was then distributed to the scorers by e-mail for review and verification. In a few cases, recording errors were discovered and corrected in subsequent scoring telecons. Recorded results were provided to the IAG for review.

4.5.3.6 IAG Review

The IAG reviewed scoring results, usually in light of preliminary processing results. The IAG asked questions, particularly regarding interpretation and application of the FOM Dictionary, and of the relationships of the scores to the related FOM. For example, if a technology was rated low in technology readiness but high in technology advancement cost (i.e. low in cost), the IAG would ask for an explanation. If the ST Lead could not provide an adequate answer or rationale to the IAG, the scores in question were revisited by the scoring team(s), and often changed.

The FOM definitions were refined based on feedback during the strawman and regular scoring sessions. As the FOM definitions were better understood over the course of the scoring and review activities, a few scores were revisited and changed.

4.5.4 Compilation and Presentation of Scoring Data

Once final scores were determined for each mission, the scoring team submitted the scores to the IAG where they were maintained in a controlled scoring database. FOM weights were applied to the scores and the resulting normalized weighted scores were plotted as a series of bar charts. The IAG assigned weights to each of the five FOM categories (Performance, Technical, Reliability/Safety, Cost and schedule) based on the primary and supporting objectives of the IISTP Phase I effort. The FOM category weights were applied to normalized weighted scores and the results plotted in the form of scatter plots. The scatter plots facilitated the cost-benefit assessment of each ISP technology for each mission.

4.5.4.1 Normalized Weighted Score Sheets

An example of the normalized scoring work sheets is given in Figure 4.5.4.1-1 for the Titan Explorer mission. For each FOM category, a normalized total was computed, based on the FOM score and weight as

$$NormalizedTotal = 100 \frac{\sum_i W_i S_i}{9 \sum_i W_i}$$

Where

W_i = weight of the ith figure of merit

S_i = score for the ith figure of merit

Note, if a technology scores the highest possible score "9" for each FOM within a FOM category, the normalized total for that technology for that FOM category is 100.

	FIGURES OF MERIT	WEIGHT	SOA Chem/ Chem	SOA Chem/ AC/Chem	SEP 5 kW/ AC/Chem	SEP 10 kW/ AC/Chem	Nuclear Electric Ion	Solar Sails/ AC/ Chem	Mag-sail (M2P2) AC/Chem
	Technology Number		1	2	10	11	13	16	17
PERFORM.	1 Payload Mass Fraction	9	1	1	9	9	9	3	9
	2 Trip Time	10	1	3	9	9	3	3	9
	4 Time on Station	0							
	19 Prop. System Launch Mass & Volume	10	1	3	3	3	1	1	9
	Normalized Total	100	11.11	26.44	77.01	77.01	46.36	25.67	100.00
TECHNICAL	3 Operational Complexity	9	9	3	3	3	9	3	3
	5 Propellant Storage Time	9	9	9	9	9	9	9	9
	6 Station Keeping Precision	0							
	14 Crew Productivity	0							
	15 Sensitivity to Malfunctions	10	3	3	3	3	3	3	3
	16 Sensitivity to Perf. Deficiencies	10	3	3	3	3	9	3	3
	17 Enable In-Space Abort Scenarios	0							
	18 Crew/Payload Exposure to In-Space Env's	0							
	Normalized Total	100	64.91	49.12	49.12	49.12	82.46	49.12	49.12
RELIABILITY/ SAFETY	31 Pre-Launch Env. Hazards & Prot.	10	1	1	1	1	1	1	1
	41 In-Space Env. Hazards & Prot.	10	3	3	3	3	1	3	3
	42 Crew Exposure & Safety	0							
	43 Payload Exposure & Protection	8	9	9	9	9	1	3	9
	51 Relative Reliability Assessment	10	9	3	3	3	1	3	3
	53 Operating Life	10	9	9	3	3	1	3	3
	Normalized Total	100	87.59	53.70	39.81	39.81	20.37	28.70	39.81
COST	61 Technology Advancement Cost	9	9	3	3	3	1	3	1
	62 Mission Non-Recurring Cost	9	9	3	3	3	1	3	3
	67 Operational Cost	10	1	1	1	1	1	1	1
	68 Mission Recurring Cost	10	9	9	3	3	1	9	9
	Normalized Total	100	76.61	45.03	33.33	33.33	16.96	50.88	45.61
SCHEDULE	81 Total Development Time	10	9	3	3	3	3	3	1
	82 Special Facility Requirements	9	9	9	9	9	9	3	3
	83 Architectural Fragility	9	9	1	3	3	3	9	9
	84 Maturity (TRL Level)	8	9	3	3	3	3	1	3
	Normalized Total	100	77.78	50.00	66.67	66.67	50.00	45.06	53.09

Figure 4.5.4.1-1 Sample Scoring Sheet for Titan Explorer

The final, weighted, consensus scores for each mission are given in the tables contained in Appendix C.

4.5.4.2 Bar-Line Plots

The tabular scoring data contained in the worksheets was graphically represented as a series of bars as shown in Figure 4.5.4.2-1 for the Titan Explorer example. The bars indicate the normalized total for each FOM category and are grouped by technology. The chart provides a quick visual depiction of how each technology scored relative to each of the FOM categories.

For example, the technologies (SOA Chemical excepted) with the best scores in each FOM category for the Titan Explorer mission are:

Performance (light-blue):   Plasma Sails (M2P2)
Technical (yellow):   NEP Ion
Reliability/Safety (green):   NTP
Cost (orange):   Solar Sails
Schedule (blue):   SEP Ion

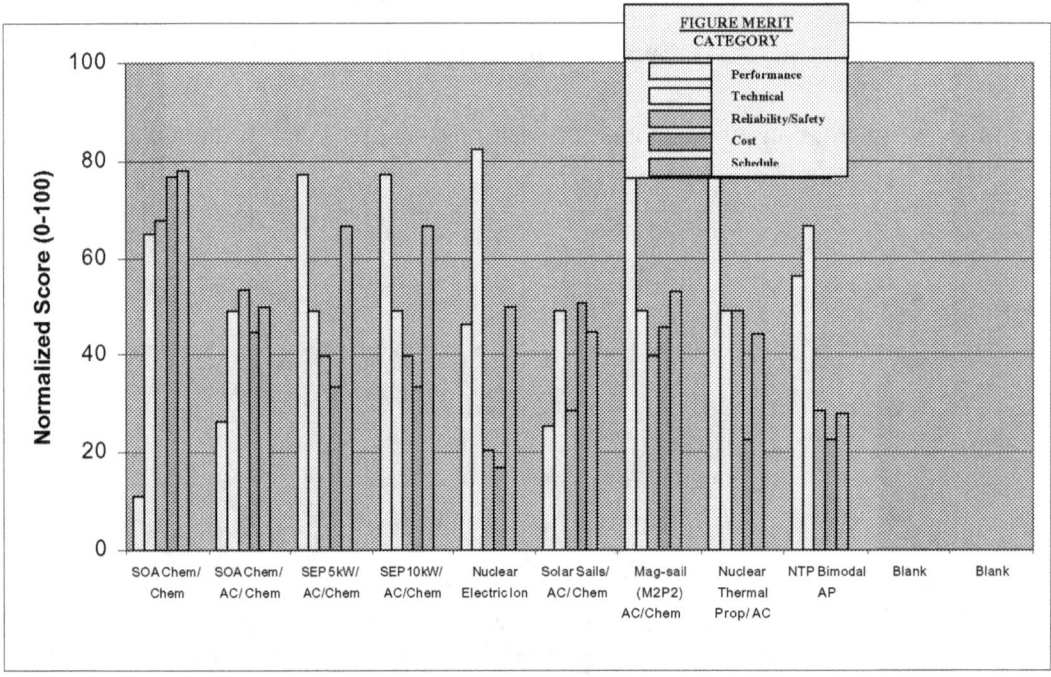

Figure 4.5.4.2-1 Parallel Bar Chart for Titan Explorer

Note that some ISP technologies may score high for some FOM categories and low for others. As an example on the Titan Explorer mission, NEP ion scored relatively high on the Technical FOM category and low on the Cost FOM category. Furthermore, NEP ion scores on the Performance and Schedule FOM categories were about average. No one technology can be expected to score the highest on all FOM categories. In addition, not all FOM categories are of equal importance to the overall goals and objectives for ISP technology selection and prioritization. Therefore, bar charts alone while providing a quick visual depiction of relative scoring, make it extremely difficult to select the best ISP candidate(s) for a given mission.

4.5.4.3 FOM Category Weights

The relative importance among the FOM categories was accounted for through the establishment and application of weights to the FOM category normalized scores. The establishment of the FOM category weights was a very important aspect of the evaluation process. In the development of any system, there are primary objectives that reflect the purpose for which the system is to be developed, and there are supporting objectives that reflect the constraints under which the system will be developed.

Specifically, the overall objective of the IISTP Phase I effort was to identify and recommend for investment those candidate ISP technologies that could most effectively and economically perform the highest priority missions. The primary objective was ISP performance; those ISP technologies that can significantly reduce trip time and increase payload mass fraction for future space missions. The supporting objectives were that the ISP system be cost-effective, safe, and reliable. In general, primary objectives support advanced technologies, while supporting objectives often support retention of current SOA technologies. Existing technologies inherently have less programmatic risk due in large part to their level of maturity and usage experience. Less programmatic risk usually results in SOA systems scoring better than advanced systems on reliability/safety, cost and schedule FOM categories. Placing too much weight on these FOM categories and on supporting objectives, would favor existing technologies, and make new technologies appear less attractive.

The IAG carefully considered FOM category weights to ensure the primary objectives and supporting objectives were properly accounted for in the final results. Performance was determined to be twice as important as cost for advance ISP technologies. Cost and Technical were equally weighed and determined to each be twice as important as either reliability/safety or schedule. The resulting FOM category weights were:

Performance	40%
Technical	20%
Reliability/Safety	10%
Cost	20%
Schedule	10%

4.5.4.4 Effectiveness versus Cost

As stated earlier, the overall objective of the IISTP Phase I effort was to identify and recommend for investment those candidate ISP technologies that could most <u>effectively</u> and <u>economically</u> perform the highest priority missions. To facilitate the evaluation of the candidate technologies based on their relative effectiveness and economies, two new parameters were defined:

Effectiveness Parameter- A measure of how well the candidate ISP technology reliably and safely performs the mission and meets the technical objectives. The Effectiveness parameter was computed by a linear combination of the normalized totals for performance, technical, and reliability/safety FOM categories and their respective relative weights and is expressed as

$$E = .57\,p + .28\,t + .14\,r$$

Where E = effectiveness parameter
p = normalized total for performance FOM category
t = normalized total for technical FOM category
r = normalized total for reliability/safety FOM category

The coefficients of .57, .28 and .14 were calculated based on the FOM category weights discussed in the previous subsection.

For example, the coefficient for p is $40/(10+20+40) = .57$.

Cost Parameter- A measure of how economical the ISP technology is in terms of cost and schedule considerations. The Cost Parameter was computed by a linear combination of the normalized total of cost and schedule FOM categories and their respective relative weights and is expressed as

$$C = .67\,c + .33\,s$$

Where C = cost parameter
c = normalized total for the cost FOM category
s = normalized total for the schedule FOM category

As with the Effectiveness Parameter, the coefficients of .67 and .33 were calculated based on the FOM category weights discussed in the previous subsection.

IISTP Phase I Final Report
September 14, 2001

The Effectiveness Parameter was plotted against the Cost Parameter for each of the nine missions, and the results are presented in Appendix C. An example of such a plot is shown in Figure 4.5.4.4-1 for the Titan Explorer mission. It is simple enough to determine how each ISP technology compares relative to the Effectiveness and Cost Parameters by looking to the top and right-most points, respectively. For the Titan Explorer, M2P2 (point #17) has the highest Effectiveness Parameter score and SOA Chemical (point #2) with aerocapture has the highest Cost Parameter score.

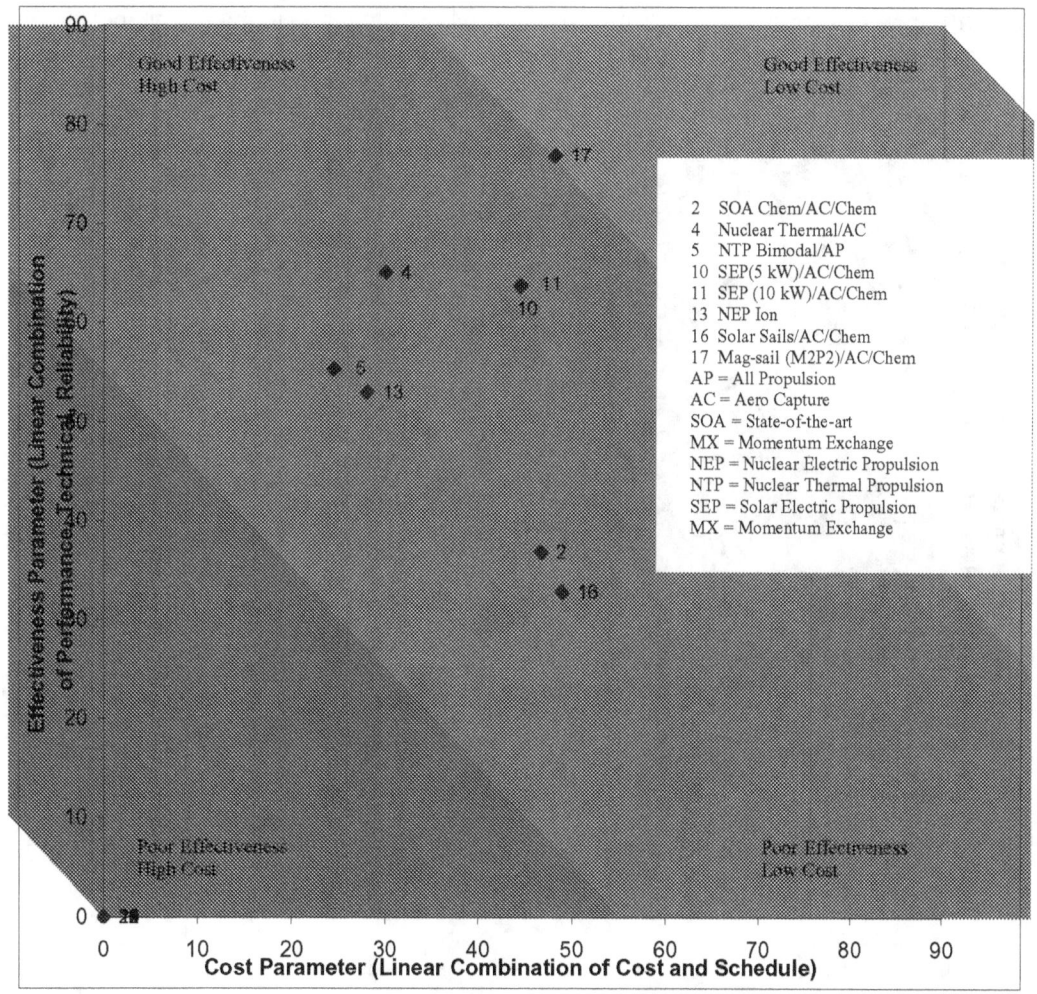

Figure 4.5.4.4-1 Cost/Effectiveness Scatter Chart - Titan Explorer Mission

To determine the ISP technology with the best combination of Effectiveness and Cost Parameter Scores, is a simple matter of overlaying a series of lines of "constant goodness" referred to as "Isos". "Isos" are lines constructed by connecting equivalent values on both the abscissa and ordinate scales as shown in Figure 4.5.4.4-2. For example, NEP Ion (point #13) and Solar Sails (point #16) lie close to the 80-80 "Iso". That is a line passing through (0, 80) and (80,0) coordinates on the ordinate and abscissa scales, respectively. Even though NEP Ion scored much better than Solar Sails on the Effectiveness Parameter score, the difference is "equally offset" by the advantage Solar Sails gained in the Cost Parameter score. For this case, both technologies are treated as competitive with one another. Similarly, the M2P2 appears to have the best overall combination (Effectiveness and Cost) score even though the SOA Chemical had a higher Cost Parameter score.

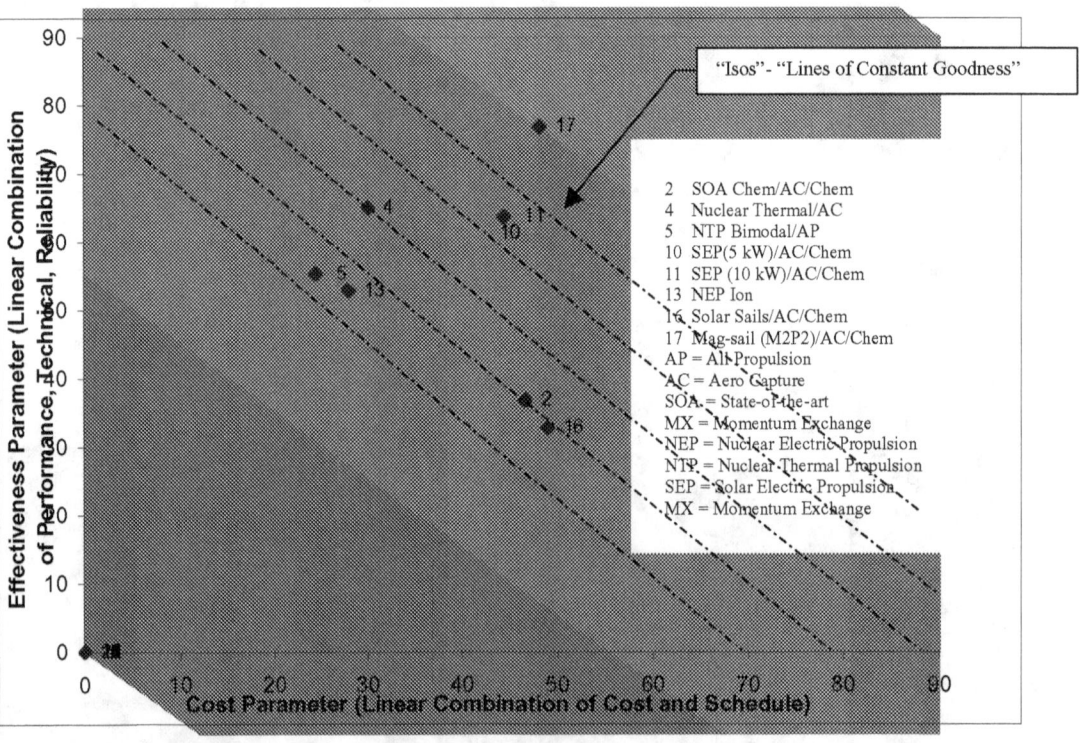

Figure 4.5.4.4-2 Cost-Effectiveness Scatter Chart for Titan Explorer Mission with "Isos" overlays

Finally, plots were generated to see how each technology compared across missions. Figure 4.5.4.4-3 shows the results for the NEP technology. "No Application" means that the technology was not applicable to the indicated missions. In this case, the plot shows that NEP systems never scored better than 40 on the Cost Parameter but had a fairly wide variation from mission to mission on the Effectiveness Parameter. This is not surprising since all NEP systems will face the same development cost and schedule challenges regardless of the mission, but be more effective on some missions than on others.

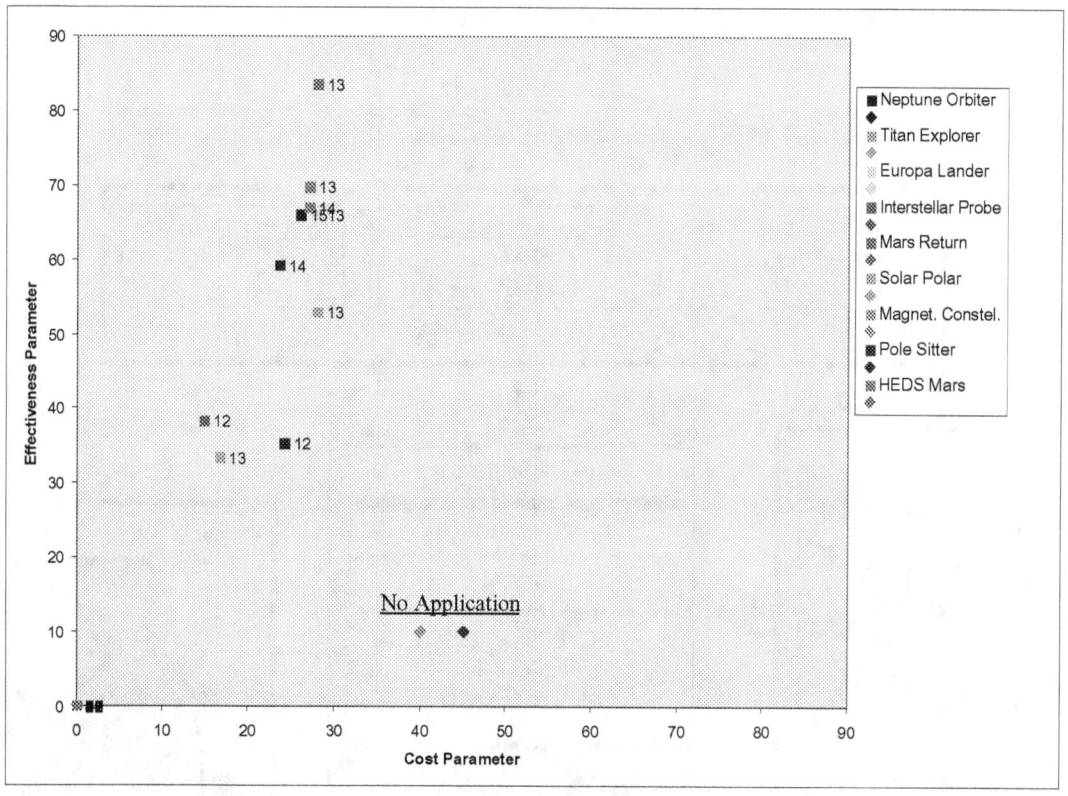

Figure 4.5.4.4-3. Cost-Effectiveness Scatter Plot for NEP for All Missions

4.6 Step 6: Prioritization of ISP Technologies

Responsibility for final prioritization of ISP technologies during Phase I was left to the IAG. During Phase I, nine missions were analyzed to evaluate more than 20 different propulsion system options against 26 FOM. The results were synthesized and represented in approximately 20 different bar-line and scatter plots. Given the extensive amount of data generated, it was decided that the most efficient way to analyze the data and formulate a set of prioritizations was to convene the IAG face-to-face in an off-site workshop. The primary objective of the workshop was to identify a prioritized set of ISP technologies that could be used to guide investment decisions. The IISTP Technology Prioritization Workshop Process is illustrated in Figure 4.6-1.

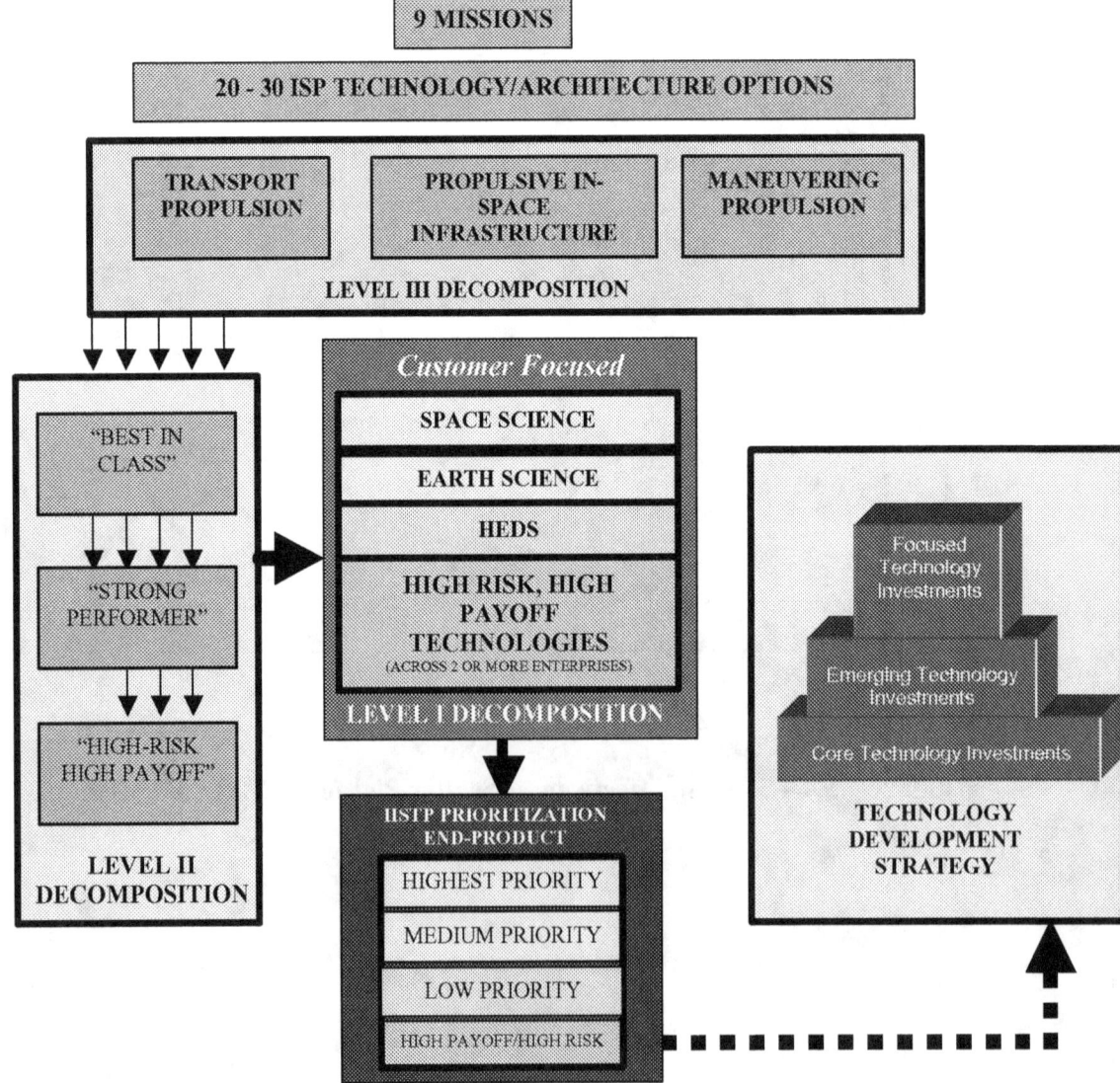

Figure 4.6-1. IISTP Technology Prioritization Workshop Process

4.6.1 Level III Decomposition

The approximate 20 propulsion systems analyzed were comprised of combinations of 17 distinctive ISP technologies. The primary objective of Level III Decomposition was to segregate the technologies into three "bins" according to their In-Space function:

1) <u>Transport</u>: ISP technologies used to transport the payload to the destination.

2) <u>Propulsive In-Space "Infrastructure"</u>: ISP technologies that provide an infrastructure in space for repetitive use on multiple missions such as momentum tethers.

3) <u>Maneuvering</u>: ISP technologies used to maneuver the payload at the destination.

The results of Level III Decomposition are given in Table 4.6.1-1. All but three of the technologies could be classified as transport technologies. Some technologies were labeled "TBA" or "to be analyzed", since it was believed more analysis was required to effectively determine their potential role mission role.

ISP TECHNOLOGY	TRANSPORT	PROPULSIVE IN-SPACE "INFRASTRUCTURE"	MANEUVERING
SOA Chemical (pivot)			
Advanced Chemical	X	TBA	TBA
NTP	X		
NTP Bimodal	X		TBA
MX Tether		X	
SEP Hall	X	X	TBA
SEP Ion	X		TBA
NEP Hall	X	TBA	TBA
NEP Ion	X		TBA
NEP VaSIMR	X		
NEP MPD	X		
Solar Sails	X	TBA	X
Plasma Sails	X	TBA	TBA
Radio-Isotope	TBA		
Solar Thermal	X	TBA	TBA
NTP/NEP Hybrid	X	TBA	TBA
Aero-Capture	X		

Table 4.6.1-1 Level III Decomposition Results

4.6.2 Level II Decomposition

Once the transport ISP technologies were segregated during Level III Decomposition, each was evaluated using the data and scores generated prior to the workshop. The primary objective of Level II Decomposition was to segregate these technologies according to how well they scored in Step 5. The scatter plots of Overall Performance versus Overall Cost measures, presented in Appendix C, were used extensively during the Level II Decomposition.

There were three scoring bins:

1) <u>Best in Class</u>: ISP technologies that scored highest on at least one of the nine missions analyzed.

2) <u>Strong Performer</u>: ISP technologies that scored well (i.e. Effectiveness Parameter score greater than 50%) over a majority of the nine missions.

3) <u>High Risk/High Payoff</u>: ISP technologies that are considered to be high risk due to their low TRL, but have a potential for high payoff should they be developed.

The results of the Level II Decomposition are given in Table 4.6.2-1.

ISP Technology	"Best In Class"	"Strong Performer"	"High Risk/High Payoff"
SOA Chemical (pivot)			
Advanced Chemical	MSR	EL	
NTP		NO,MSR,TE,MP	
NTP Bimodal		MSR,MP	
MX Tethers	NO,EL		X
SEP Hall	MP, MC		
SEP Ion	NO, TE, MSR,EL,PS	MC,SPI	
NEP Hall			
NEP Ion		MSR, MP, NO, TE, EL	
NEP VaSIMR		NO	
NEP MPD		NO++	
Solar Sails	SPI, PS		
Solar Sails (1gm/m2)	ISP		X
Plasma Sails	NO,ISP,MSR,EL,TE		X
Solar Thermal		MC	
NTP/NEP Hybrid			
Aero-Capture	NO,MSR, TE, MP		

KEY: NO- Neptune Orbiter EL- Europa Lander MP- Mars Piloted TE- Titan Explorer PS- Pole Sitter SPI- Solar Polar Imager MC- Magnetospheric Constellation ISP- Interstellar Probe MSR- Mars Sample Return X- High Risk/High Payoff

Table 4.6.2-1 Level II Decomposition Results

4.6.3 Level I Decomposition

The primary objective of the Level I Decomposition activity was to determine each of the Code Y, S, and M Enterprise priorities. Each of the Enterprise customers was asked to rate the technologies for their respective Enterprises based on High, Medium, Low, and High Risk/High Payoff.

The results of the Level I Decomposition are given in Table 4.6.3-1.

ISP Technology	HIGH	MEDIUM	LOW	HIGH PAYOFF/HIGH RISK
Advanced Chemical		S, M		
NTP		M	S	
NTP Bimodal			M	
MX Tethers				S,M
SEP Hall	M			
SEP Ion	S			
NEP Hall			M	
NEP Ion		S		
NEP VaSIMR				M
NEP MPD	M			
Solar Sails	S		M	
Solar Sails (1gm/m2)				
Plasma Sails				S,M
Solar Thermal		M		
NTP/NEP Hybrid				
Aero-Capture	S,M			
Precision Station Keeping Placeholder	S,Y			

S = Code S Priority
M = Code M Priority
Y = Code Y Priority

Table 4.6.3-1 Level I Decomposition Results

4.6.4 IISTP Technology Prioritization End-Product

The final step, in the IISTP Phase I Workshop process, was to combine all of the results into a cross-Enterprise prioritized set of ISP technologies that could be used to guide investment decisions. The IAG as a whole reached a consensus and the results are given in Table 4.6.4-1.

HIGH	MEDIUM	LOW	HIGH PAYOFF/ HIGH RISK	DROP
Aerocapture (for robotic & HEDS)	Solar Sails	Solar Thermal	Plasma Sail	NTP
SEP Ion (5/10 kW)	SEP Hall (100kW)	Bi-modal NTP (Low- to High-Power Scalable)	MXER Tether	
NEP (Low- to High-Power Scalable)	Class I Electric Propulsion (30kW - 100kW 3,000 - 10,000 sec eff >50%)		Solar Sail (1gm/m2)	
	Advanced Chemical (cryo +TBD)			
	Class II Electric Propulsion (> 500kW > 3000 sec eff >50%)			

Table 4.6.4-1 IISTP Phase I Consensus Results

5.0 Conclusions

ISP technologies were identified and prioritized according to their relative payoff and their ability to perform and/or enable customer prioritized missions. ISP technologies were selected based on their ability to effectively and economically support multiple NASA Enterprises. All applicable advanced ISP technologies were analyzed on at least one mission.

The evaluation, selection and prioritization process was designed to ensure the candidate ISP technologies maximized mission success and minimized mission risks. The evaluation, selection and prioritization process provided the rationale and data needed to guide investment decisions.

The IISTP Phase I effort concluded with a consensus across NASA Programs, Projects, Technology Centers and Enterprises as to those technologies that deserve consideration in future investment decisions.

6.0 Recommendations

For a fixed vehicle mass in low earth orbit (LEO), the vehicle mass may be allocated between propellant and payload mass. This yields a trade space of trip time and payload mass. One of the benefits of advanced propulsion systems is that they provide a relatively large trip time/payload mass trade space when compared with conventional chemical propulsion systems. In this study, time and resources did not permit quantitative definition of this trade space. It is recommended that future studies quantitatively define the trip time/payload mass trade space for several selected missions for several selected advanced propulsion technologies. It is further recommended that customers examine the trade space and make recommendations on optimal selection of trip time and payload mass within the trade space.

For each combination of propulsion technology and mission, there is an optimal specific impulse. This is derived from the fact that as specific impulse increases, propellant mass decreases, but the fixed propulsion mass required to achieve the specific impulse increases. Therefore, there is usually some specific impulse for which the total mass (dry mass plus propellant mass) is a minimum. It is recommended that future efforts include the definition of the relationship between specific impulse and propulsion system mass for selected combinations of propulsion system technology and mission to enable propulsion system developers to focus their efforts on specific impulse ranges most beneficial to future NASA missions.

For most of the figures of merit, the 0, 1, 3, 9 scoring system representing degrees to which the propulsion technology satisfied the figure of merit was adequate. This is particularly true for figures of merit for which the degree of satisfaction of the figure of merit is substantially qualitative. However, there are some figures of merit for which quantitative analysis is possible and appropriate. Trip time and payload mass fraction are two important figures of merit for which quantitative definition is both possible and appropriate. The detailed derivation of quantitative figures of merit requires more resources than were available for the present study. However, it is recommended that in future studies, quantitative analysis be used to define these parameters quantitatively for selected combinations of propulsion technologies and missions. These results may be used as adjuncts to the type of analysis used in the present study or may be used as direct scores in the scoring process.

In the scoring process, scores were established after verbal interchange among the systems team or other scoring unit. Usually, the rationale behind the score was lost in the process. It is recommended that in future efforts, a brief rationale behind the scores be recorded on a systematic basis for review by others. It is further recommended that the schedule for scoring be modified to accommodate the recording of rationale.

There are two program needs that must be addressed in the future. First, there is a need to formulate quantitative results where appropriate, address the trip time/payload mass trade space, and address the issue of optimal specific impulse as discussed above. Second, the state of the knowledge and the state of development of the various advanced propulsion technologies changes over time. The state of definition of missions changes over time, and new missions are conceived over time. Furthermore, it is hoped and anticipated that mission planners will conceive new missions enabled by the ISP program and/or modify currently planned missions based on new mission capabilities such as availability of large quantities of electric power at the destination. Therefore, as time progresses, there is a need to repeat this process and make appropriate changes in recommendations as knowledge about the propulsion technologies and of the missions changes. It is recommended that the issues of quantitative results, trip time/payload mass trade space, and optimal specific impulse be addressed in FY02 for several selected combinations of propulsion technology and missions that have been addressed in this study. Furthermore, it is recommended that the process described in this report be repeated in FY03 and on a bi-annual basis thereafter and that the selection of technologies be reviewed on a bi-annual basis until advanced technologies are sufficiently developed that they can be definitively assigned to missions.

7.0 Acronyms and Abbreviations

AHP - Analytic Hierarchy Process

ASTP – Advance Space Technology Program

AU – Astronomical Unit

Code M - Human Exploration and Development of Space Enterprise

Code S – Space Science Enterprise

Code Y – Earth Science Enterprise

CNSR - Comet Nucleus Sample Return

CT – Cost Team

DS1 – Deep Space 1

EASI – Earth Atmospheric Solar Occultation Imager

ED – Electrodynamic

EP – Electronic Propulsion

EREMF – Earth Radiative Energy Measurement Facility

ESA – European Space Agency

FOM – Figure of Merit

GEC - Geospace Electrodynamic Connections

GEO – Geosynchronous Earth Orbit

GRC – Glenn Research Center, Cleveland, Ohio

GSFC – Goddard Space Flight Center

HEDS – Human Exploration and Development of Space

HEDM – High Energy Density Matter

IAG - IISTP Advisory Group

IISTP – Integrated In-Space Transportation Plan

ISP – In-Space Propulsion

ISPP - In-Situ Propellant Production

ITAC – Integrated Technology Assessment Center

JSC – Johnson Space Center, Houston, TX

JPL – Jet Propulsion Laboratory

LaRC – Langley Research Center

LEO – Low Earth Orbit

M2P2 - Mini-Magnetospheric Plasma Propulsion

MPD - magnetoplasmadynamic

MRT – Mission Requirements Team

MSFC – Marshall Space Flight Center, Huntsville, AL

MX – Momentum Exchange

NASA – National Aeronautics and Space Administration

NEP – Nuclear Electric Propulsion

NGST – Next Generation Space Telescope

NSTAR - NASA Solar Electric Technology Application Readiness

NTP – Nuclear Thermal Propulsion

NTR – Nuclear Thermal Rocket

QFD - Quality Function Deployment

OMB – Office of Management and Budget

OSS – Office of Space Science

PER - overall performance measure

POC – Point of Contact

PIT – Pulsed Inductive Thruster

PRO - overall programmatic measures

R_E – Earth Radii

SEP – Solar Electric Propulsion

SIM - Space Interferometry Mission

SOA – State-of-the-Art

ST – Systems Team

TBA – To Be Analyzed

TRL – Technology Readiness Level

TT - Technology Team

T/W – Thrust-to-Weight Ratio

VASIMR - Variable Specific Impulse Magnetoplasma Rocket

REFERENCES

1. http://www.jpl.nasa.gov/adv_tech/ballutes/misn_neptune.htm
2. http://www.jpl.nasa.gov/adv_tech/ballutes/misn_titan.htm
3. http://sse.jpl.nasa.gov/site/missions/B/europa_lander_network.html
4. http://www.jpl.nasa.gov/adv_tech/ballutes/misn_mars.htm
5. http://science.nasa.gov/ssl/pad/solar/suess/Interstellar_Probe/ISP-Intro.html
6. http://umbra.nascom.nasa.gov/spd/secr/missions/polarimg.html
7. http://stp.gsfc.nasa.gov/missions/mc/mc.htm
8. http://lws.gsfc.nasa.gov/lws_resources_imagegallery.htm
9. http://nssdc.gsfc.nasa.gov/planetary/mars/mars_crew.html
10. http://siliconsky.com/sao/fit/nuclear.htm
11. http://www.tethers.com/
12. http://spacsun.rice.edu/aspl/vasimr.htm
13. http://www.howstuffworks.com/solarsail.htm
14. http://spike.geophys.washington.edu/Space/SpaceModel/M2P2/
15. http://www.msfc.nasa.gov/STD/propulsion/research/solar/
16. http://siliconsky.com/sao/fit/electric.htm
17. Kepner and Tregoe; "The Rational Manager", 2001.
18. Kepner and Tregoe; "Decision Making in the Digital Age", 2001.
19. Akoa, "Quality Function Deployment: Integrating Customer Requirements into Product Design", 1990.
20. Mizuno and Akao, "QFD: The Customer-Driven Approach to Quality Planning and Deployment ", 1994.
21. Saaty, T.L. "The Analytic Hierarchy Process.", 1980.
22. Saaty, T.L. "Decision Making for Leaders", 1996.
23. Saaty, T.L. "Deriving Bayes Theorem from the Analytic Hierarchy Process", 1992.

APPENDIX A

IISTP Team Rosters

Systems Team

Name	Office
Joe Bonometti	MFSC Systems, Solar Thermal
Bob Cataldo	GRC Power Systems
Bret Drake	JSC Systems, Human Missions
Len Dudzinski	GRC Systems, NTP/NEP, Trajectories/Sizing, Fusion
Robert Frisbee	JPL Systems, Sails
Leon Gefert	GRC/Lead, Systems POC
Jeff George	JSC/Lead, Systems, NEP
Rob Hoyt	TU Tethers, Sizing
Jonathan Jones	MSFC Plasma Sails, Technology Team Lead
Larry Kos	MSFC/Lead, Chemical/Trajectories/Sizing
Melissa McGuire	GRC NTP Systems, Trajectories/Sizing
Jim Moore	SRS Systems, ED Tethers, STP
Michelle Munk	LaRC/Lead, Aeroassist
Mahmoud Naderi	MSFC Cost
Muriel Noca	JPL/Lead, Systems, & Team-X POC, Sizing
Tara Poston	MSFC Trajectories/Sizing
Bob Sefcik	GRC Cost
Kirk Sorensen	MSFC MX Tethers
Nobie Stone	SRS Systems, ED Tethers, STP
Gordon Woodcock	Gray Research/ITAC
Scott Baird	JSC ISPP Systems
John Blandino	JPL POC for Code S
Neil Dennehy	GSFC POC
Sandy Kirkindall	MFSC Systems
Saroj Patel	MFSC Systems

Consultants for ST: Juan Aone, Chen-Wan Yen (JPL Sail & EP Trajectories), Steve Oleson (GRC SEP data), Steve Tucker, Dave Plachta (MSFC & GRC CFM)

IISTP Phase I Final Report
September 14, 2001

Technology Team

Represented by a POC for each ASTP Technology Element
and appropriate research areas

Name	Office	Propulsion Category/Type
		Electric Propulsion
Mike Patterson	GRC	Ion
Rob Jankovsky	GRC	Hall
Franklin Chang Diaz	JSC	VASIMR
Mike LaPoint	Ohio Aerospace Institute	MPD gas-fed
Jay Polk	JPL	MPD lithium-fed
Mike LaPoint	Ohio Aerospace Institute	PIT
Scott Benson	GRC	SEP Systems
Hoppy Price	JPL	Solar Sails
Jonathan Jones	MSFC	Plasma Sails
Stan Borowski	GRC	Fission (NTR)
Mike Houts	MSFC	Fission (NEP)
Bob Estes	Harvard Smithsonian	Electrodynamic Tethers
Kirk Sorensen	MSFC	Momentum Exchange Tethers
Michelle Munk	LaRC	Aeroassist
Scott Baird	JSC	In-Situ Propellant Production
Bill Taylor	GRC	Advanced Chemical
Don Bai	MSFC	Advanced Chemical
Hartwell Long	JPL	Advanced Chemical
Jeff Weiss	JPL	Light Weight Components
Steve Tucker	MSFC	Cryogenic Fluid Management
Joe Bonnemeti	MSFC	Solar Thermal Propulsion
Scott Benson	GRC	Pulsed Plasma Thruster

Consultant for TT: Gordon Woodcock, Gray Research

IISTP Advisory Group

Name	Office
Harley Thronson	Space Science IISTP POC
Loren Lemmerman	Earth Science IISTP POC
Richard Fischer	Exploration and Development POC
Rae Ann Meyer	MSFC, Space Transfer Technologies Assistant Project Manager
Randy Baggett	MSFC, Propellantless Propulsion Project Manager
Harry Cikanek	GRC POC
Tim O'Donnell	JPL POC
Larry Kos	MSFC, IISTP Systems Analysis Lead
Jonathan Jones	MSFC, IISTP Technology Lead

Consultants for IAG: Gray Research - Deborah Sims, Bob Farris, Bill Eberle, Gordon Woodcock

Mission Requirements Team

Team	Name	Technology/Office
Space Science Team	Harley Thronson	Code S Lead[1]
Earth Science Team	Loren Lemmerman	Code Y Lead
	John Lebreque	Code Y Science Theme[2]
	Eduardo Torres	Code Y Visions and Decadal Planning
Exploration and Development	Richard Fischer	Code M Lead[3]

[1]**Representing each Office of Space Science Theme**
Solar System Exploration
Sun Earth Connection
Astronomical Search for Origins
Structure and Evolution of the Universe

[2]**Representing these Earth Science Themes**
Atmospheric Chemistry
GWEC
Oceans and Ice

[3]**Representing these major Offices of Space Flight Areas/Programs**
Human Exploration
Commercialization and Development of Space
International Space Station
Space Shuttle
Space Operations Management Office

Cost Team

Name	Office
Mahmoud Naderi	MSFC Cost
Sharon Czarnecki	SAIC
Robert Sefcik	GRC
Gordon Woodcock	Gray research

IISTP Phase I Final Report
September 14, 2001

APPENDIX B

Figures of Merit Dictionary

National Aeronautics and Space Administration
George C. Marshall Space Flight Center

IISTP FOM Dictionary
Revision E.14
April 30, 2001

Figures Of Merit Dictionary

For

Integrated In-Space Transportation Plan (IISTP)

Propulsion Technology Evaluation

**ADVANCED SPACE TRANSPORTATION PROGRAM (ASTP) PROGRAM OFFICE
TD15**

APPROVALS/CONCURRENCES

Prepared By:

Approved By:

_____
Bob Farris
Consultant
Manager
Gray Research Inc.

_____
Les Johnson
In-Space Transportation Investment Area

NASA Marshall Space Flight Center

Concurrence By:

_____
Richard Fischer
Code M Advanced Programs Office
NASA Headquarters

_____
Mahmoud Naderi
IISTP Cost Team Lead
NASA Marshall Space Flight Center

_____
Harley Thronson
Code S Technology Director (Acting)
NASA Headquarters

_____
Jonathan Jones
IISTP Technology Team Lead

_____
Loren Lemmerman
Code Y Lead
NASA Jet Propulsion Laboratory

_____
Larry Kos
IISTP Systems Team Lead
NASA Marshall Space Flight Center

_____
Tim O'Donnell
JPL POC
NASA Jet Propulsion Laboratory

_____
Joe Nainiger
GRC POC
NASA Glenn Research Center

REVISIONS

REV	DATE	AUTHOR	FOM	DESCRIPTION
A	3-23-01	Farris	18	Changed to include payload exposure
A	3-23-01	Farris	43	Changed to include spacecraft exposure and protection
A	3-23-01	Farris	44	Combined with FOM 43
A	3-23-01	Farris	51	**Changed From:** MTBF **To:** Relative Reliability Assessment
A	3-23-01	Farris	52	Combined with FOM 51
A	3-23-01	Farris	86	**Changed From:** Requires Exotic Materials **To:** Requires Exotic Materials and/or Processes
A	3-23-01	Farris	87	Combined with FOM 86
A	3-23-01	Farris	88	Combined with FOM 86
A	3-23-01	Eberle	4	Changed to describe station keeping functions which will be included
A	3-23-01	Eberle	6	Included formation flying in description
A	3-23-01	Eberle	68	Add clarifying test to last line of first paragraph
A	3-23-01	Eberle	3	Grammar
B	4-12-01	Farris	Rev E	Not released/All changes reflected in revisions to Rev A in Rev E below
C	4-12-01	Farris	Rev E	Not released/All changes reflected in revisions to Rev A in Rev E below
D	4-12-01	Farris	Rev E	Not released/All changes reflected in revisions to Rev A in Rev E below

REVISIONS (continued)

E	4-12-01	Kos	53	**Changed From:** "Systems operating life…" **To:** "Systems operating life (versus the demonstrable operating life)…" (In two places)
E	4-12-01	Kos		(1) Created new FOM Category titled "Applicability" (2) Moved FOMs 63,64, 66,69,70 **From:** Cost category 　**To:** Applicability Category (3) Changed Scoring Responsibility **From:** Cost Team 　**To:** Systems Team
E	4-12-01	Kos	65	Deleted FOM 65 titled "Special Handling Requirements"
E	4-12-01	Kos	69	Changed name **FROM:** Reduce # of Missions Required to Support Initial Mission **TO:** Missions Required to Support Initial Mission
E	4-12-01	Kos	70	Changed name **FROM:** Reduce # of Missions Required to Support Follow-On Missions **TO:** Missions Required to Support Follow-On Missions
E	4-12-01	Farris	31	Change Title **From:** "Pre-Launch Environmental Hazards/Protection" **To:** "Ground Operations Environmental Hazards/Protection" Include FOM 32 and 33
E	4-12-01	Farris	32,33	Deleted and Incorporated into FOM 31
E	4-12-01	Farris	51,53	Change Scoring Responsibility **From:** Systems Team **To:** Technology Team
E	4-12-01	Vane	19	Created new FOM titled "Total Propulsion System Launch Mass and Volume"
E	4-12-01	Farris	44,52, 87,88	Removed all FOMs titled "RESERVED"
E	4-12-01	Vane	81	Changed Title **From:** "Development Time" **To:** "Total Development Time" Revised definition accordingly to include FOM 86.
E	4-16-01	Vane	86	Deleted and Incorporated into FOM 81
E	4-18-01	Kos	13	Deleted and Incorporated into FOM 63
E	4-18-01	Kos	63	Changed Title **From:** "Applicability" **To:** "Applicability/Adaptability/Flexibility" Revised definition accordingly to include FOM 13.
E	4-18-01	Naderi	67	**Added words:** "…including launch vehicle purchase…" to Operational Cost FOM
E	4-18-01	Naderi	68	**Added words:** "…(less launch vehicle cost which is to be included in Operational Cost FOM #67)…" to Mission Recurring Cost FOM
E	4-18-01	Kos	85	Deleted. During Neptune Orbiter evaluation there were no technologies found which had significant embedded new technologies. This FOM can in general be considered a part of FOM 84
E	4-20-01	Johnson	1,2,6, 19	Create new major FOM category titled "Performance" and moved FOM 1, 2, 6, and 19 into the new category.
E	4-20-01	Johnson	84	Added rationale for a score of "0" to that used for a score of "1", to ensure that no ISP will receive a score of "0" based on this FOM.

IISTP FOM Dictionary
Revision E.14
April 30, 2001

1.0 PURPOSE

The purpose of this document is to define the figures of merit (FOMs) to be used in the initial phase of the In-Space Propulsion (ISP) technology prioritization effort. It is not the intention or purpose of this document to identify and define a comprehensive set of FOMs for ISP technologies. Rather, the goal is to identify and define a relatively concise set of measures that adequately support the ISP technology prioritization process. These FOMs will be used to evaluate candidate ISP technologies. These FOMs were provided and agreed to by the Codes S, Y, and M Leads. These FOMs were selected based on knowledge of the candidate ISP technologies and mission categories for which the candidate technologies may be used.

Sections 2.0 – 7.0 are used to define each of the FOMs. Guidelines and responsibilities for ISP technology scoring are contained within each of the FOM definitions. The FOMs are grouped according to six major categories:

   2.0 Performance
   3.0 Technical
   4.0 Reliability/Safety
   5.0 Cost
   6.0 Applicability
   7.0 Schedule

Appendix A is used to document the weights for each of the FOMs that have been provided by Codes Y, S and M leads according to their respective mission categories. Weights are maintained in a separate Appendix to the main FOM Dictionary document with access limited only to those organizations not directly or indirectly involved in the ISP technology evaluations and scoring.

2.0 PERFORMANCE DEFINITIONS

Figure of Merit #	FOM Category	Scoring Responsibility	Title
1	*Performance*	*Systems Team*	*Payload Mass Fraction*

This measure is the ratio of payload mass at the destination to total vehicle mass in low earth orbit. Payload mass is the total mass required to achieve scientific objectives at the destination. Neither the inert propulsion system mass nor the propellant mass required to reach the destination is included in payload mass. If a power source is used at the destination and is also used for propulsion to reach the destination then that part of the propulsion system mass which is used for power generation at the destination may be included in payload mass. This figure of merit is coupled to the trip time; therefore scoring should be done in tandem (see Figure of Merit #2).

Figure of Merit #	FOM Category	Scoring Responsibility	Title
2	*Performance*	*Systems Team*	*Trip Time*

This is the total transportation time required to achieve all scientific objectives of the mission. If the mission includes a crew return or a sample return, then trip time includes the time required for return to earth. In general, the trip time includes only the time required for transportation functions. For example, for a human trip to Mars or a Mars sample return, the trip time would include only the transportation time and would exclude time spent on the Martian surface.

It is recognized that there is a trade between payload mass fraction and trip time. Available mass may be divided between payload mass or additional propellant to decrease trip time. Therefore, the payload mass fraction and trip time figures of merit are related. A propulsion technology which provides for the best values of both payload mass fraction and trip time should be rated as a "9" in both categories. A propulsion technology which provides only a marginal payload mass fraction and a poor trip time should be rated a "1" in both categories. A propulsion technology which provides a good value of payload mass fraction or trip time without significant sacrifice to the either should be rated a "3" in both categories.

Figure of Merit #	FOM Category	Scoring Responsibility	Title
6	Performance	Systems Team	Station Keeping Precision

This is a measure of the ability of the propulsion technology to perform precision station keeping or formation flying functions. In general, it is a function of the minimum impulse capability of the propulsion technology. The propulsion technology with the smallest minimum impulse bit capability shall receive a score of "9". Other propulsion technologies shall receive scores of "3" or "1", depending on whether their minimum impulse bit is somewhat greater than or significantly greater than that of the best propulsion technology.

Station keeping and formation flying will be considered in this phase of the ISP technology evaluation only if they require a significant ΔV (e.g., pole sitting missions). Station keeping and formation flying will not be considered in this phase if they are used for minor adjustments in Keplerian motion. Future phases of the evaluation will consider this issue.

Figure of Merit #	FOM Category	Scoring Responsibility	Title
19	Performance	Systems Team	Total Propulsion System Launch Mass and Volume

This is a relative measure of the total propulsion system launch mass and volume. Those ISP technologies with less total propulsion system launch mass will enable smaller launch vehicles to be used for a given mission. Similarly, ISP technologies that can be packaged in a small volume can be launched aboard a wider variety of vehicles.

ISP technologies with relatively small propulsion system launch mass and volume requirements shall receive a score of "9". ISP technologies with relatively large propulsion system launch mass and/or volume requirements shall receive a score of "1". All remaining ISP technologies shall receive a score of "3".

3.0 TECHNICAL DEFINITIONS

Figure of Merit #	FOM Category	Scoring Responsibility	Title
3	Technical	Systems Team	Operational Complexity

Operational complexity relates to the number, sensitivity, and complexity of the propulsion-related operations that must be performed during the mission. An ISP technology which requires a large number of complex operations should be rated a "1". An ISP technology which may require a significant number of operations should be rated a "3", if the operations are relatively simple. An ISP technology which requires only a few simple operations should be rated a "9".

Figure of Merit #	FOM Category	Scoring Responsibility	Title
4	Technical	Systems Team	Time on Station

Several missions require station keeping or formation flying. Station keeping is defined as keeping the vehicle in a required orientation or in a required position relative to the planetary body of interest at the destination. Formation flying is keeping the vehicles of a constellation in correct position and/or orientation with respect to each other. This criterion measures the ability of the vehicle to maintain time on station (as limited by available propellant) relative to the time required for the vehicle to perform its required scientific function. The propulsion technology which provides the capability to remain in position at the destination for the greatest amount of time shall receive a score of "9". Other propulsion technologies shall receive scores of "3" or "1", depending on how they perform relative to the best propulsion technology.

Station keeping and formation flying will be considered in this phase of the ISP technology evaluation only if they require a significant ΔV (e.g., pole sitting missions). Station keeping and formation flying will not be considered in this phase if they are used for minor adjustments in Keplerian motion. Future phases of the evaluation will consider this issue.

Figure of Merit #	FOM Category	Scoring Responsibility	Title
5	Technical	Systems Team	Propellant Storage Time

This criterion is a measure of the degree to which propellant storage is an issue for the mission. ISP technologies in which propellant storage is not an issue (e.g. propellantless technologies, technologies that use easily storable propellants, or technologies that have a sufficiently short flight time) shall receive a score of "9". ISP technologies which use propellants which are inherently difficult to store but lend themselves to simple, reliable storage solutions for long duration missions shall receive a score of "3". ISP technologies which use propellants which are difficult to store (e.g., liquid hydrogen) and storage solutions become an issue even for moderately long missions shall receive a score of "1".

Figure of Merit #	FOM Category	Scoring Responsibility	Title
14	Technical	Systems Team	Crew Productivity

This is a relative measure of crew productivity losses associated with the presence of the candidate ISP technology. Scores should reflect a composite of the estimated times required for the crew to maintain and/or operate the candidate ISP technology. ISP technologies which require no extraordinary involvement or interaction by the crew shall receive the highest score of "9". ISP technologies in which crew maintenance and operation intervals are relatively short, simple, and predictable (i.e., can be scheduled) and require little crew involvement/interaction shall receive a score of "3". Those ISP technologies that could potentially require extensive crew interaction/involvement for long periods of time and/or at unexpected times during the mission shall receive a score of "1".

Figure of Merit #	FOM Category	Scoring Responsibility	Title
15	Technical	Systems Team	Sensitivity to Malfunctions

This is a relative measure of how the candidate ISP technology malfunctions or fails, the relative consequences of the malfunction or failure and the extent of the techniques required to minimize the consequences of a malfunction or failure. ISP technologies whose most likely failure modes would not result in a loss of life, a significant loss of property or a significant loss of mission objectives shall receive a score of "9". ISP technologies whose most likely failure modes, unmitigated, would result in a grave loss of property, life or mission and/or require extensive and complex methods for mitigation shall receive a score of "1". Those ISP technologies whose most likely failure modes while potentially significant can be mitigated with simple well-known methods shall receive a score of "3".

Figure of Merit #	FOM Category	Scoring Responsibility	Title
16	Technical	Systems Team	Sensitivity to Performance Deficiencies

This is a measure of the relative consequences of a performance deficiency or "shortfall". A performance deficiency is different that a malfunction or failure. A performance deficiency occurs when the operational or in-mission performance is less than the predicted and/or previously tested performance. Some technologies can inherently recover from an unforeseen deficiency without consequences to the mission; these ISP technologies shall receive a score of "9". Other technologies have inherent performance margins and can recover in time without severe consequences to the mission; these ISP technologies shall receive a score of "3". Other technologies have little or no margin for performance deficiencies; "shortfalls" would pose a significant risk to the mission. For these technologies, increasing performance margin would result in definite increases in cost and/or weight. These ISP technologies shall receive a score of "1".

Figure of Merit #	FOM Category	Scoring Responsibility	Title
17	*Technical*	*Systems Team*	*Enable In-Space Abort Scenarios*

This is a relative measure of how well the candidate ISP technology supports in-space aborts and a safe return to earth for human missions. In addition, this measure should be used to evaluate the ability of the candidate technology to support new and innovative abort scenarios. An ISP technology that permits rapid return to earth in the event of an emergency (related or unrelated to the propulsion system) and can easily accommodate new and innovative abort scenarios shall be rated a "9". An ISP technology that supports rapid return to earth but cannot easily accommodate new and innovative abort scenarios shall be rated a "3". An ISP technology which can support in-space aborts but with great difficulty shall be rated a "1".

Figure of Merit #	FOM Category	Scoring Responsibility	Title
18	*Technical*	*Systems Team*	*Crew &/or Payload Exposure to In-Space Environments*

This is a measure used to assess the degree to which the candidate ISP technology can be used to minimize crew and/or payload exposure to adverse natural environments. Examples include long-term exposure to zero-g or natural radiation because of long trip times or long residence times in the earth's radiation belts, respectively. An ISP technology that can effectively minimize crew and/or payload exposure to in-space environments with minimal changes to the mission profile shall be rated a "9". An ISP technology that can effectively minimize crew and/or payload exposure to in-space environments but requires significant changes in the mission profile shall be rated a "3". An ISP technology that provides only modest reductions in crew and/or payload exposure to in-space environments and requires significant changes in the mission profile shall be rated a "1". An ISP technology that inherently (by the very nature of its operations) increases crew and/or payload exposure to in-space environments shall be rated a "0".

4.0 RELIABILITY/SAFETY DEFINITIONS

Figure of Merit #	FOM Category	Scoring Responsibility	Title
31	Reliability/ Safety	Systems Team	Ground Operations Environmental Hazards and Protection

This is a relative measure of the hazards the candidate ISP technology poses to the environment, ground crew and ground support equipment during ground operations. ISP technologies which pose no significant risk or impact to the environment, ground crew and ground support equipment by the nature of their operation, handling or materials on the ground while maintaining, enhancing or providing for new ground test abort options shall be scored a "9". Those ISP technologies which (1) could potentially create grave and/or irreversible damage to the environment; (2) create grave and/or lethal harm to the ground crew; (2) could potentially result in extreme loss or damage to the ground support equipment; (3) severely restrict ground test abort options; and/or (4) require extensive and complex methods for mitigation; shall receive a score of "1". Those ISP technologies whose threat to the environment can be mitigated with simple well-known methods shall receive a score of "3".

Figure of Merit #	FOM Category	Scoring Responsibility	Title
41	Reliability/ Safety	Systems Team	In-Space Environmental Hazards and Protection

This is a relative measure of the hazards the candidate ISP technology poses to the in-space environment during the mission. ISP technologies which pose no significant risk or impact to the in-space environment by the nature of their operation, handling or materials during the mission shall be scored a "9". Those ISP technologies which could potentially create grave and/or irreversible damage to the in-space environment and require extensive and complex methods for mitigation shall receive a score of "1". Those ISP technologies whose threat to the in-space environment can be mitigated with simple well-known methods shall receive a score of "3".

Figure of Merit #	FOM Category	Scoring Responsibility	Title
42	Reliability/ Safety	Systems Team	Crew Exposure and Safety

This is a relative measure of the hazards the candidate ISP technology poses to the crew during the mission. ISP technologies which pose no significant risk or injury to the crew by the nature of their operation, handling or materials while maintaining, enhancing or providing for new abort options shall be scored a "9". Those ISP technologies which could potentially create grave and/or lethal harm to the crew, severely restricting abort options and require extensive and complex methods for mitigation shall receive a score of "1". Those ISP technologies whose threat to the crew can be mitigated with simple well-known methods shall receive a score of "3".

Figure of Merit #	FOM Category	Scoring Responsibility	Title
43	*Reliability/ Safety*	*Systems Team*	*Payload and/or Spacecraft Exposure and Protection*

This is a relative measure of the hazards (including optics/detector exposure to contamination) the candidate ISP technology poses to the payload and/or spacecraft during the mission. ISP technologies which pose no significant risk or impact to the payload and/or spacecraft performance by the nature of their operation, handling or materials during the mission shall be scored a "9". Those ISP technologies which could potentially result in extreme loss or damage to the payload and/or spacecraft and require extensive and complex methods for mitigation shall receive a score of "1". Those ISP technologies whose threat to the payload and/or spacecraft can be mitigated with simple well-known methods shall receive a score of "3".

Figure of Merit #	FOM Category	Scoring Responsibility	Title
51	*Reliability/ Safety*	*Technology Team*	*Relative Reliability Assessment*

This is a measure of the relative reliability expected from the candidate ISP technology during its required operating life.

The relative reliability assessment can be made based on the number, type, complexity, typical failure modes, failure rates and time to recover for the critical ISP elements/components. Typically, those technologies with the fewest, simplest, most reliable components with the greatest amount of redundancy should be scored a "9". ISP technologies with the largest number of complex, fragile, and intricate components with the greatest number of single point failures should be scored a "1". All other ISP technologies should be scored a "3".

Figure of Merit #	FOM Category	Scoring Responsibility	Title
53	*Reliability/ Safety*	*Technology Team*	*Operating Life*

System operating life (versus the demonstrable operating life) can be determined by estimating operating life of the individual parts, components, subsystems and systems and of the candidate ISP technology as a whole. Should operating life (versus the demonstrable operating life) be a mission requirement, operating life must be quantified to the extent necessary to estimate operating life margin for each candidate ISP technology.

If operating life is not a hard mission requirement, then relative estimates of operating life can be made based on the number, type, complexity and typical operating life of the critical components used in each of the candidate ISP technologies. Typically, those technologies with the fewest, simplest components with the longest estimated operating life should be scored a "9". ISP technologies with the largest number of complex, fragile, and intricate components with the shortest estimated operating life should be scored a "1". All other ISP technologies should be scored a "3".

IISTP FOM Dictionary
Revision E.14
April 30, 2001

5.0 COST DEFINITIONS

Figure of Merit #	FOM Category	Scoring Responsibility	Title
61	Cost	Cost Team	Technology Advancement Cost

This measure includes costs for advancing the technology from its current state of development to TRL 6 or 7, including special test facilities if facility modifications or new facilities are required. The decision of appropriate level (6 vs 7) will be made for each technology according to the "technology acceptance criteria" determinations. If a technology flight demonstration is deemed necessary (TRL 7) this is included in technology advancement.

ISP technologies which can reach the desired technology level based on "business as usual" funding in Code R receive a score of "9". Those which require a moderate amount of special augmentation funding receive a score of "3". Those which appear to require establishment of a dedicated program and Congressional line item receive a score of "1".

Figure of Merit #	FOM Category	Scoring Responsibility	Title
62	Cost	Cost Team	Mission Non-Recurring Cost

Development costs begin with program definition studies aimed at creating a formal procurement specification for the in-space propulsion system and end with completion of a successful qualification test program that demonstrates satisfaction of the specification. Any new test and/or production facilities are included in this cost. For purposes of consistency, we will assume that the development program produces no flight units. If a developmental mission, such as a New Millennium mission, is deemed necessary, that will be priced as a mission (not development) and will be assumed to use the first production unit.

ISP technologies for which the non-recurring cost is deemed to be within the funding capability of the first customer mission receive a score of "9". ISP technologies which require a special development program (i.e. are beyond the capability of first user to fund) receive a score of "3". ISP technologies which appear to require so much development funding as to impact the overall NASA budget receive a score of "1".

Figure of Merit #	FOM Category	Scoring Responsibility	Title
67	Cost	Cost Team	Operational Cost

This measure includes all of the costs associated with operating the propulsion system, beginning with ground processing at the launch site, and including launch vehicle purchase as well as mission/flight operations. For practical purposes, these are lumped in with mission operations costs, but costs of extra personnel and facilities associated with operation of the propulsion system should be considered in estimating mission operations costs. For example, if the candidate ISP technology involves nuclear subsystems, there may be costs associated with extra range safety efforts, and nuclear specialists may be required on mission operations staff.

If the ISP technology results in less than 10% of the overall mission operations cost the candidate technology receives a score of "9". If the overall mission operations cost exceeds 10% but does not require special staffing or analytical methods/procedures to support launch or mission operations the candidate technology receives a score of "3". If the ISP technology requires special staffing or analytical methods/procedures, or if it creates a significant operations cost burden compared to the pivot, it receives a score of "1".

Figure of Merit #	FOM Category	Scoring Responsibility	Title
68	Cost	Cost Team	Mission Recurring Cost

This is a measure of the total cost (less launch vehicle cost which is to be included in Operational Cost FOM #67), priced by manufactured unit, for production of flight-ready in-space propulsion systems, and integration with the spacecraft payload. This cost includes delivery to the vehicle manufacturer or to the launch site, as appropriate. If an acceptance test is required of the ISP, its cost is included in this.

If this cost is less than the launch vehicle cost, and reduces total tranportation cost relative to the pivot, it receives a score of "9". If the cost is more than the launch vehicle cost but enables a total tranpsortation cost (including launch vehicle) not more than the pivot, it receives a score of of "3". If the total transportation cost is more than the pivot, it receives a score of "1".

IISTP FOM Dictionary
Revision E.14
April 30, 2001

6.0 APPLICABILITY DEFINITIONS

Figure of Merit #	FOM Category	Scoring Responsibility	Title
63	*Applicability*	*Systems Team*	*Applicability/Adaptability/Flexibility*

This is a relative measure of the applicability/adaptability/flexibility of the candidate ISP technology for other mission categories with different performance requirements. It is assumed that as propulsion technologies are developed, certain components (e.g., nuclear power generators and electric thrusters) will be developed for a discrete set of values for significant propulsion characteristics such as design maximum power and design maximum thrust. For these types of components, it is unlikely that systems will be developed specifically for the needs of each individual mission. Rather, a set of systems will be available for "off-the shelf" application to specific missions. Requirements for each individual mission may be met by a combination of operating the system at less than its design maximum power or thrust and by combining individual systems to achieve power/thrust levels greater than the design power/thrust levels of one unit. For other components (e.g., tanks) total impulse or ΔV appropriate for a specific mission can be achieved by appropriate tank sizes for that mission.

An ISP technology which can be adapted with relative ease to satisfy a wide range of specific performance level needs (power, thrust, total impulse, ΔV, duration, duty cycle, etc.) associated with 7 or 8 mission categories receives a score of "9". A technology can be adapted to satisfy performance levels of 4 - 6 mission categories receives a score of "3". A technology which can be adapted to satisfy performance levels of 2 or 3 mission categories receives a score of "1". A technology which cannot be adapted and is useful for only a single mission category receives a score of "0".

Figure of Merit #	FOM Category	Scoring Responsibility	Title
64	*Applicability*	*Systems Team*	*Scalability (robotic - human)*

Scalability is defined as accommodating the same underlying technology, design approach and operational methods over a wide range of size, specifically from robotic missions (typical payload 500 kg) to human (typical payload 10,000+ kg).

A technology which can be scaled using the same materials, design and test approach, and operational methods receives a score of "9". If scaling <u>requires</u> changes in any one of these, the technology is scored "3". If 2 or all are changed, it receives a score of "1". If it cannot be reasonably scaled to human missions, it receives a score of "0". If a mission application chooses to change any of these, that is not counted against the technology. (For example, a human mission may choose to cluster engines.)

Figure of Merit #	FOM Category	Scoring Responsibility	Title
66	*Applicability*	*Systems Team*	*Supports Evolutionary Development of a Long-Term Capability*

Evolutionary development indicates the technology supports creation of permanent or long-lasting in-space transportation infrastructure which can be used to support continuing missions.

If the technology supports creation of an infrastructure which supports missions over a decade or more without replacement, it receives a score of "9". If it requires partial replacement, or supports missions over a period of 5 years to a decade, it receives a score of "3". If the infrastructure is relatively temporary but supports more than one mission, the technology receives a score of "1". If no infrastructure or repeat mission capability exists, it receives a score of "0".

Figure of Merit #	FOM Category	Scoring Responsibility	Title
69	*Applicability*	*Systems Team*	*Missions Required To Support Initial Mission*

This figure of merit applies to ISPs or missions where support launches or missions may be required prior to the initial "objective" mission (i.e. the mission that accomplishes program or mission objectives). Examples are emplacement of a momentum-exchange tether facility (ISP) or pre-placement of cargo on Mars prior to the first human landing (mission). The objective is to minimize these. An ISP technology that requires no prior missions receives a score of "9" (for example, it might have enough performance to deliver Mars cargo along with crew). An ISP technology for which the cost of the prior mission(s) plus the first objective mission is less than for the pivot ISP technology receives a score of "3" unless it qualifies for the "9". An ISP technology for which the cost of the prior mission(s) plus the first objective mission is more than the pivot ISP receives a score of "1".

Figure of Merit #	FOM Category	Scoring Responsibility	Title
70	*Applicability*	*Systems Team*	*Missions Required To Support Follow-On Missions*

This figure of merit applies to ISPs or missions where support launches or missions may be required prior to continuing "objective" missions. Examples are replenishment of an orbital facility or pre-placement of additional cargo on Mars prior to continuing human landings. The objective is to minimize these. An ISP technology that requires no prior missions receives a score of "9" (for example, it might have enough performance to deliver Mars cargo along with crew). An ISP technology for which the cost of the prior mission(s) plus the supported objective mission is less than for the pivot ISP technology receives a score of "3" unless it qualifies for the "9". An ISP technology for which the cost of the prior mission(s) plus the supported objective mission is more than the pivot ISP technology receives a score of "1".

7.0 SCHEDULE DEFINITIONS

Figure of Merit #	FOM Category	Scoring Responsibility	Title
81	Schedule	Technology Team	Total Development Time

This figure of merit applies to the total estimated technology advancement and full-scale development time required for the candidate ISP technology to be developed and ready to support a specific mission. The total time should be obtainable from the technology roadmap. This FOM includes technology advancement to TRL 6, any flight demonstrations or development tests required, need for exotic materials, assembly, and/or testing, materials availability, materials characterization, special testing facility requirements, life testing of hardware, and full-scale development time from ATP to completion of qualification. For purposes of evaluation, assume no funding gaps in the activity. (Exotic materials are those not available by routine commercial order from normal industrial suppliers, such as special-order alloys.)

Any ISP technology at TRL 6 with a "nominal" time estimated for full-scale development shall receive a score of "9". "Nominal" means no additional time is expected for construction of facilities, flight demonstrations, lengthy life tests, exotic materials, or exotic assembly. Full-scale development of a chemical rocket engine would be a typical "nominal" schedule.

Any ISP technology shall receive a score of "3" when the estimated total development time adds 4 or fewer years to a "nominal" schedule. These ISP technologies may require exotic materials, however, the characteristics data is assumed to be readily available from the supplier or the technical literature. These ISP technologies may require exotic materials, however, the lead times are estimated to be less than a year. These ISP technologies may require exotic assembly processes, however, any assembly required should be able to be accomplished in an existing facility.

Any ISP technology shall receive a score of "1" when the estimated total development time adds 5 or more years to a "nominal" schedule. These ISP technologies require materials for which data are not readily available or the data are uncertain, and whose lead times are greater than a year. These ISP technologies require exotic assembly techniques or tests using new special facilities or life testing of more than 2 years.

Figure of Merit #	FOM Category	Scoring Responsibility	Title
82	Schedule	Technology Team	Special Facility Requirements

This figure of merit concerns special facility requirements for testing, production, or launch/recovery and support operations. In-space infrastructure and assembly are covered by separate figures of merit.

An ISP technology with no special facility requirements, or can use existing, readily available facilities (that is there are no associated schedule issues), and the cost is nominal, receives a score of "9". An ISP technology with modest special facility construction requirements, or when scheduling an existing special facilities schedule delays can be held to 1 year or less, receives a score of "3". An ISP technology requiring construction of major new special facilities, or use of existing special facilities which pose severe scheduling problems, receives a score of "1".

Figure of Merit #	FOM Category	Scoring Responsibility	Title
83	Schedule	Technology Team	Architectural Fragility

This figure of merit refers to potential problems commonly associated with project eventualities that result in a major change in the architecture. For example, the Apollo mission architecture was fragile because if the spacecraft mass had grown slightly more than it did (or engine performance had fallen short), the mission would not have been possible on a single launch and no backup plan existed. To make this evaluation, mission architecture must be associated with the ISP technology.

An ISP technology/architecture which can readily adapt to project eventualities such as weight growth, power requirements growth, or moderate subsystem performance shortfalls (e.g. by altering mission design or adding launches) receives a score of "9". An ISP technology/architecture that can adapt by selecting an available larger launch vehicle receives a score of "3". An ISP technology/architecture that must be redesigned to adapt receives a score of "1". An ISP technology/ architecture that cannot adapt receives a score of "0".

Figure of Merit #	FOM Category	Scoring Responsibility	Title
84	Schedule	Technology Team	Maturity (TRL Level)

This figure of merit applies to current technology status.

An ISP technology that is at TRL 6, or will reach TRL 6 under current program plans and funding within 2 years receives a score of "9". An ISP technology that is at TRL 4 or 5 and can be expected to reach TRL 6 under current program plans, part of which are not currently funded, within 4 years, receives a score of "3". An ISP technology less advanced than this for which the schedule for advancement to TRL 6 cannot be forecast due to major technical uncertainties receives a score of "1".

APPENDIX C

Scores and Results

		FIGURES OF MERIT	WEIGHT	SOA Chem/ Chem	SOA Chem/ AC/ Chem	Adv.Chem/ Chem.	Nuclear Thermal Prop/ AC	NTP Bimodal AP	MX Tether/ Augmen- tation AP	MX Tether/ Augment. AC/ Chem
		Technology Number		1	2	3	4	5	6	7
PERFORM.	1	Payload Mass Fraction	9	1	1	1	9	3	3	9
	2	Trip Time	10	1	9	1	9	3	3	9
	4	Time on Station	0							
	19	Prop. System Launch Mass & Volume	10	1	3	3	3	3	9	9
		Normalized Total	100	11.11	49.43	18.77	77.01	33.33	56.32	100.00
TECHNICAL	3	Operational Complexity	9	9	3	9	3	9	1	1
	5	Propellant Storage Time	9	9	9	3	9	3	9	9
	6	Station Keeping Precision	0							
	14	Crew Productivity	0							
	15	Sensitivity to Malfunctions	10	3	3	3	3	3	3	3
	16	Sensitivity to Perf. Deficiencies	10	9	3	9	3	9	3	3
	17	Enable In-Space Abort Scenarios	0							
	18	Crew/Payload Exposure to In-Space Env's	0							
		Normalized Total	100	82.46	49.12	66.67	49.12	66.67	43.86	43.86
RELIABILITY/ SAFETY	31	Pre-Launch Env. Hazards & Prot.	10	1	1	1	1	3	1	1
	41	In-Space Env. Hazards & Prot.	10	3	3	3	1	1	3	3
	42	Crew Exposure & Safety	0							
	43	Payload Exposure & Protection	8	3	9	9	9	3	9	9
	51	Relative Reliability Assessment	10	9	3	9	3	3	1	1
	53	Operating Life	10	9	9	9	9	3	9	9
		Normalized Total	100	56.48	53.70	67.59	49.07	28.70	49.07	49.07
COST	61	Technology Advancement Cost	9	9	3	9	1	1	1	1
	62	Mission Non-Recurring Cost	9	9	3	9	1	1	1	1
	67	Operational Cost	10	1	1	1	3	3	3	1
	68	Mission Recurring Cost	10	9	9	9	3	3	9	9
		Normalized Total	100	76.61	45.03	76.61	22.81	22.81	40.35	34.50
SCHEDULE	81	Total Development Time	10	9	3	3	3	3	1	1
	82	Special Facility Requirements	9	9	9	9	1	1	9	9
	83	Architectural Fragility	9	1	3	3	9	3	9	9
	84	Maturity (TRL Level)	8	9	3	3	3	3	1	1
		Normalized Total	100	77.78	50.00	50.00	44.44	27.78	55.56	55.56

Figure C-1. Neptune Orbiter Weights and Scores (1 of 3)

	FIGURES OF MERIT	WEIGHT	SEP Hall/ NTP/ AC/Chem	SEP Hall/Chem/ AC/Chem	SEP 5 kW/ AC/Chem	SEP 10 kW/ AC/Chem	NEP Hall/ Chem/ AC/Chem	Nuclear Electric Ion	Nuclear Electric VaSIMR
	Technology Number		8	9	10	11	12	13	14
PERFORM.	1 Payload Mass Fraction	9	9	9	9	9	9	9	9
	2 Trip Time	10	3	1	9	9	1	9	9
	4 Time on Station	0							
	19 Prop. System Launch Mass & Volume	10	1	3	3	3	1	1	1
	Normalized Total	100	46.36	46.36	77.01	77.01	38.70	69.35	69.35
TECHNICAL	3 Operational Complexity	9	1	1	3	3	1	9	9
	5 Propellant Storage Time	9	3	9	9	9	9	9	3
	6 Station Keeping Precision	0							
	14 Crew Productivity	0							
	15 Sensitivity to Malfunctions	10	1	1	3	3	1	3	1
	16 Sensitivity to Perf. Deficiencies	10	3	3	3	3	3	9	9
	17 Enable In-Space Abort Scenarios	0							
	18 Crew/Payload Exposure to In-Space Env's	0							
	Normalized Total	100	22.22	38.01	49.12	49.12	38.01	82.46	60.82
RELIABILITY/ SAFETY	31 Pre-Launch Env. Hazards & Prot.	10	1	1	1	1	1	3	3
	41 In-Space Env. Hazards & Prot.	10	1	3	3	3	1	1	1
	42 Crew Exposure & Safety	0							
	43 Payload Exposure & Protection	6	1	3	9	9	1	1	1
	51 Relative Reliability Assessment	10	1	1	3	3	1	3	1
	53 Operating Life	10	3	3	3	3	3	1	1
	Normalized Total	100	15.74	24.07	39.81	39.81	15.74	20.37	15.74
COST	61 Technology Advancement Cost	9	1	3	3	3	1	1	1
	62 Mission Non-Recurring Cost	9	1	3	3	3	1	1	1
	67 Operational Cost	10	1	3	3	3	1	3	3
	68 Mission Recurring Cost	10	1	3	3	3	1	1	1
	Normalized Total	100	11.11	33.33	33.33	33.33	11.11	16.96	16.96
SCHEDULE	81 Total Development Time	10	3	3	3	3	3	3	1
	82 Special Facility Requirements	9	1	9	9	9	3	3	3
	83 Architectural Fragility	9	9	9	9	9	9	9	9
	84 Maturity (TRL Level)	8	3	3	3	3	3	3	0
	Normalized Total	100	44.44	66.67	66.67	66.67	50.00	50.00	36.42

Figure C-1. Neptune Orbiter Weights and Scores (2 of 3)

		FIGURES OF MERIT	WEIGHT	Nuclear Electric MPD	Solar Sails/ AC/ Chem	Mag-sail (M2P2) AC/Chem	Mag-Sail (M2P2) AP	Radio-isotope Electric	NTP/NEP Hybrid	Solar Thermal Prop/ AC
		Technology Number		15	16	17	18	19	20	21
PERFORM.	1	Payload Mass Fraction	9	9	3	9	9	9	3	3
	2	Trip Time	10	9	9	9	3	9	3	3
	4	Time on Station	0							
	19	Prop. System Launch Mass & Volume	10	1	1	9	9	3	1	1
		Normalized Total	100	69.35	48.66	100.00	77.01	77.01	25.67	25.67
TECHNICAL	3	Operational Complexity	9	9	3	3	9	3	3	1
	5	Propellant Storage Time	9	9	9	9	9	9	9	1
	6	Station Keeping Precision	0							
	14	Crew Productivity	0							
	15	Sensitivity to Malfunctions	10	3	3	3	3	3	3	3
	16	Sensitivity to Perf. Deficiencies	10	9	3	3	9	3	9	3
	17	Enable In-Space Abort Scenarios	0							
	18	Crew/Payload Exposure to In-Space Env's	0							
		Normalized Total	100	82.46	49.12	49.12	82.46	49.12	66.67	22.81
RELIABILITY/ SAFETY	31	Pre-Launch Env. Hazards & Prot	10	3	1	1	1	1	3	1
	41	In-Space Env. Hazards & Prot	10	1	3	3	3	3	1	3
	42	Crew Exposure & Safety	0							
	43	Payload Exposure & Protection	8	1	3	9	3	9	3	3
	51	Relative Reliability Assessment	10	3	3	3	9	3	1	1
	53	Operating Life	10	1	3	3	3	3	1	3
		Normalized Total	100	20.37	28.70	39.81	42.59	39.81	19.44	24.07
COST	61	Technology Advancement Cost	9	1	3	1	1	3	1	3
	62	Mission Non-Recurring Cost	9	1	3	3	3	3	1	3
	67	Operational Cost	10	3	3	3	3	3	3	3
	68	Mission Recurring Cost	10	1	9	9	9	3	1	3
		Normalized Total	100	16.96	50.88	45.61	45.61	33.33	16.96	33.33
SCHEDULE	81	Total Development Time	10	1	3	1	1	3	3	3
	82	Special Facility Requirements	9	3	3	9	9	9	1	9
	83	Architectural Fragility	9	9	9	9	9	9	9	9
	84	Maturity (TRL Level)	8	3	1	0	0	3	3	3
		Normalized Total	100	43.83	45.06	53.09	53.09	66.67	44.44	50.00

Figure C-1. Neptune Orbiter Weights and Scores (3 of 3)

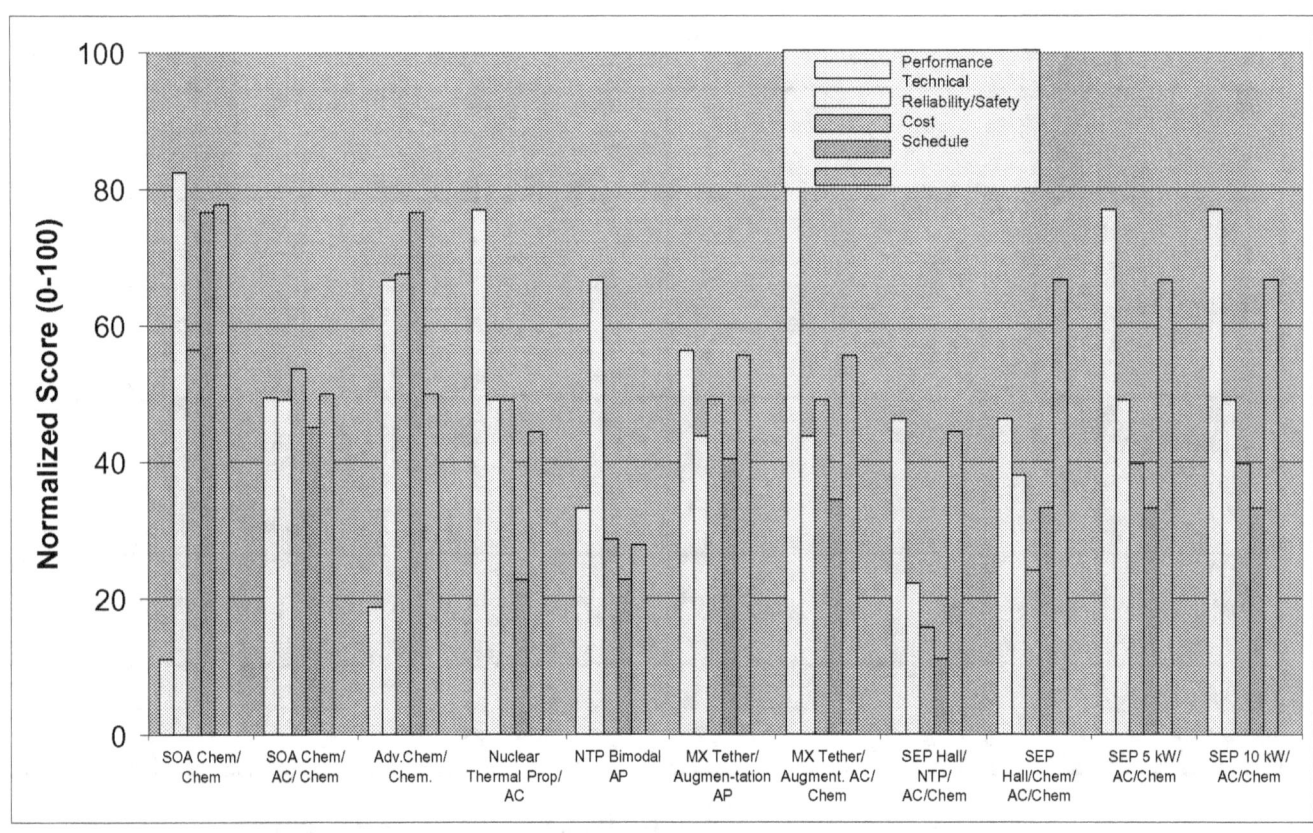

Figure C-2. Neptune Orbiter-- Normalized Scores by Major Evaluation Area (1 of 2)

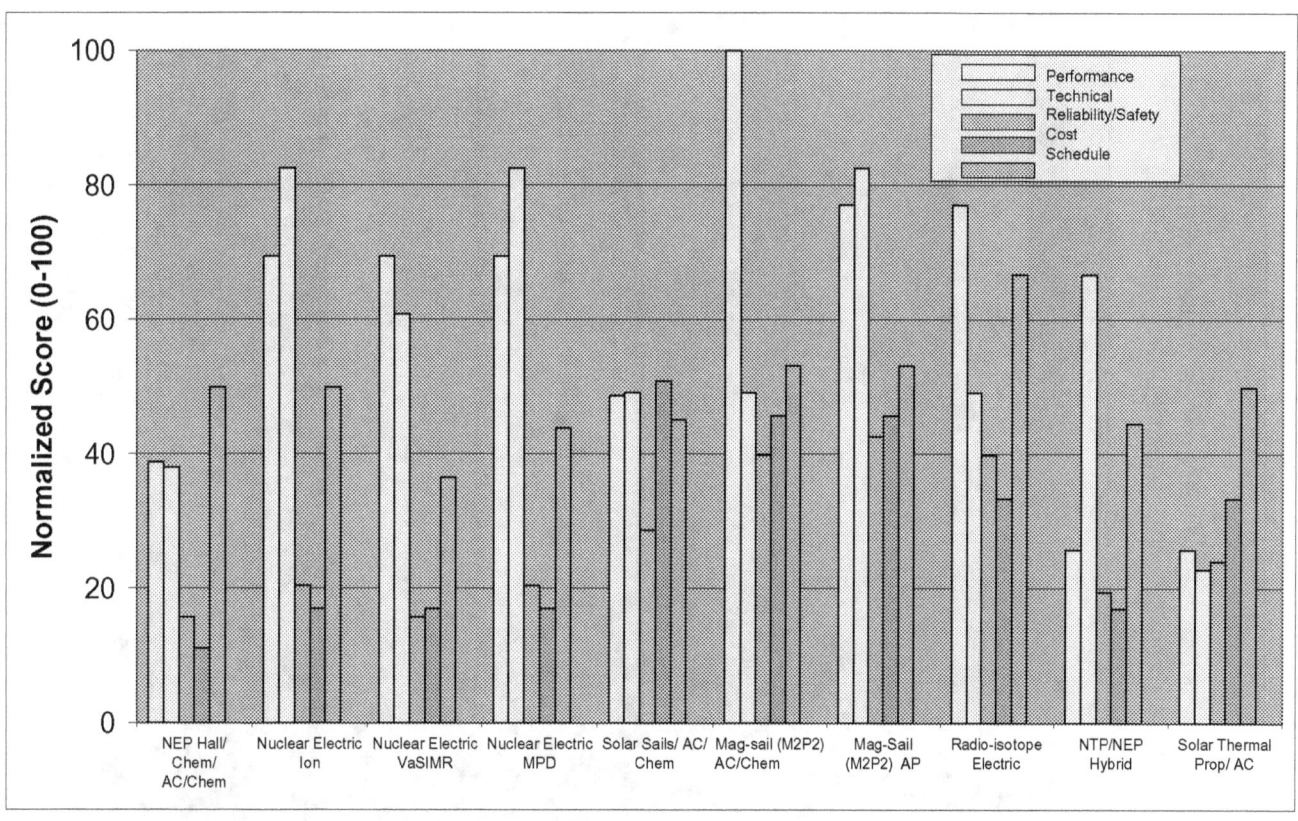

Figure C-2. Neptune Orbiter--Normalized Scores by Major Evaluation Area (2 of 2)

IISTP Phase I Final Report
September 14, 2001

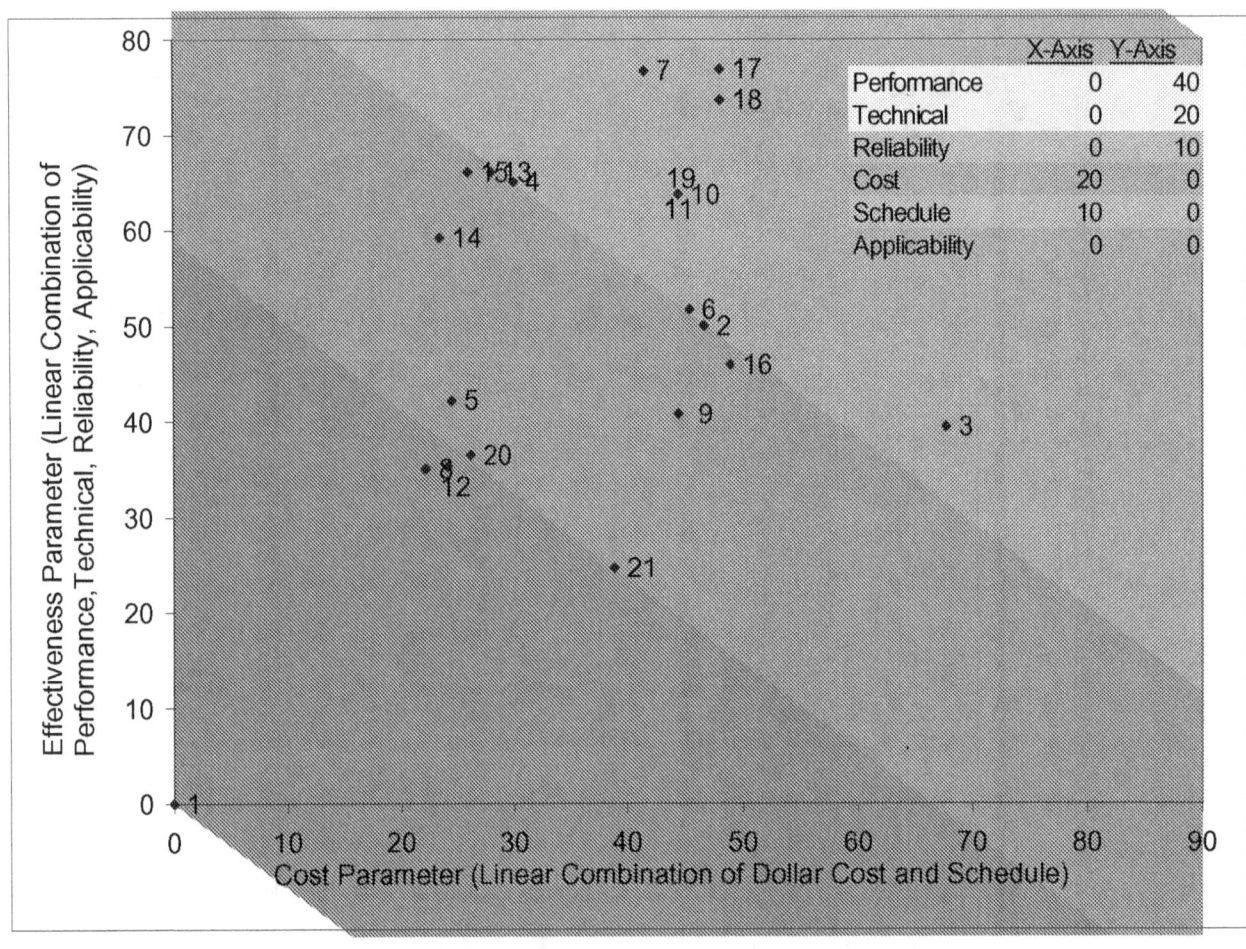

Figure C-3. Neptune Orbiter Cost-Effectiveness Scatter Chart

1 Not used
2 SOA Chem/AC/Chem
3 Adv. Chem/Chem
4 Nuclear Thermal/AC
5 NTP Bimodal/AP
6 MX Tether/Augment/AP
7 MX Tether/Augment/AC/Chem
8 SEP Hall/NTP/AC/Chem
9 SEP Hall/Chem/AC/Chm
10 SEP(5 kW)/AC/Chem
11 SEP (10 kW)/AC/Chem
12 NEP Hall/Chem/AC/Chem
13 NEP Ion
14 NEP VaSIMR
15 NEP MPD
16 Solar Sails/AC/Chem
17 Mag-sail (M2P2)/AC/Chem
18 Mag-sail (M2P2) AP
19 Radio-isotope Electric
20 NTP/NEP Hybrid
21 Solar Thermal Prop./AC

AP = All Propulsion
AC = Aero Capture
SOA = State-of-the-art
MX = Momentum Exchange
NEP = Nuclear Electric Propulsion
NTP = Nuclear Thermal Propulsion
SEP = Solar Electric Propulsion
MX = Momentum Exchange

FIGURES OF MERIT	WEIGHT	SOA Chem/ Chem	SOA Chem/ AC/ Chem	SEP 5 kW/ AC/Chem	SEP 10 kW/ AC/Chem	Nuclear Electric Ion	Solar Sails/ AC/ Chem
Technology Number		1	2	10	11	13	16
1 Payload Mass Fraction	9	1	1	9	9	9	3
2 Trip Time	10	1	3	9	9	3	3
4 Time on Station	0						
19 Prop. System Launch Mass & Volume	10	1	3	3	3	1	1
Normalized Total	100	11.11	26.44	77.01	77.01	46.36	25.67
3 Operational Complexity	9	9	3	3	3	9	3
5 Propellant Storage Time	9	9	9	9	9	9	9
6 Station Keeping Precision	0						
14 Crew Productivity	0						
15 Sensitivity to Malfunctions	10	3	3	3	3	3	3
16 Sensitivity to Perf. Deficiencies	10	3	3	3	3	9	3
17 Enable In-Space Abort Scenarios	0						
18 Crew/Payload Exposure to In-Space Env's	0						
Normalized Total	100	64.91	49.12	49.12	49.12	82.46	49.12
31 Pre-Launch Env. Hazards & Prot.	10	1	1	1	1	3	1
41 In-Space Env. Hazards & Prot.	10	3	3	3	3	1	3
42 Crew Exposure & Safety	0						
43 Payload Exposure & Protection	8	9	9	9	9	1	3
51 Relative Reliability Assessment	10	9	3	3	3	3	3
53 Operating Life	10	9	9	3	3	1	3
Normalized Total	100	67.59	53.70	39.81	39.81	20.37	28.70
61 Technology Advancement Cost	9	9	3	3	3	1	3
62 Mission Non-Recurring Cost	9	9	3	3	3	1	3
67 Operational Cost	10	1	1	3	3	3	3
68 Mission Recurring Cost	10	9	9	3	3	1	9
Normalized Total	100	76.61	45.03	33.33	33.33	16.96	50.88
81 Total Development Time	10	9	3	3	3	3	3
82 Special Facility Requirements	9	9	9	9	9	3	3
83 Architectural Fragility	9	1	3	9	9	9	9
84 Maturity (TRL Level)	8	9	3	3	3	3	1
Normalized Total	100	77.78	50.00	66.67	66.67	50.00	45.06

Figure C-4. Titan Explorer - Weights and Scores (1 of 2)

	FIGURES OF MERIT	WEIGHT	Mag-sail (M2P2) AC/Chem	Nuclear Thermal Prop/ AC	NTP Bimodal AP
	Technology Number		17	4	5
PERFORM.	1 Payload Mass Fraction	9	9	9	3
	2 Trip Time	10	9	9	9
	4 Time on Station	0			
	19 Prop. System Launch Mass & Volume	10	9	3	3
	Normalized Total	100	100.00	77.01	56.32
TECHNICAL	3 Operational Complexity	9	3	3	9
	5 Propellant Storage Time	9	9	9	3
	6 Station Keeping Precision	0			
	14 Crew Productivity	0			
	15 Sensitivity to Malfunctions	10	3	3	3
	16 Sensitivity to Perf. Deficiencies	10	3	3	9
	17 Enable In-Space Abort Scenarios	0			
	18 Crew/Payload Exposure to In-Space Env's	0			
	Normalized Total	100	49.12	49.12	66.67
RELIABILITY/ SAFETY	31 Pre-Launch Env. Hazards & Prot.	10	1	1	3
	41 In-Space Env. Hazards & Prot.	10	3	1	1
	42 Crew Exposure & Safety	0			
	43 Payload Exposure & Protection	8	9	9	3
	51 Relative Reliability Assessment	10	3	3	3
	53 Operating Life	10	3	9	3
	Normalized Total	100	39.81	49.07	28.70
COST	61 Technology Advancement Cost	9	1	1	1
	62 Mission Non-Recurring Cost	9	3	1	1
	67 Operational Cost	10	3	3	3
	68 Mission Recurring Cost	10	9	3	3
	Normalized Total	100	45.61	22.91	22.81
SCHEDULE	81 Total Development Time	10	1	3	3
	82 Special Facility Requirements	9	9	1	1
	83 Architectural Fragility	9	9	9	3
	84 Maturity (TRL Level)	8	0	3	3
	Normalized Total	100	53.09	44.44	27.78

Figure C-4. Titan Explorer - Weights and Scores (2 of 2)

IISTP Phase I Final Report
September 14, 2001

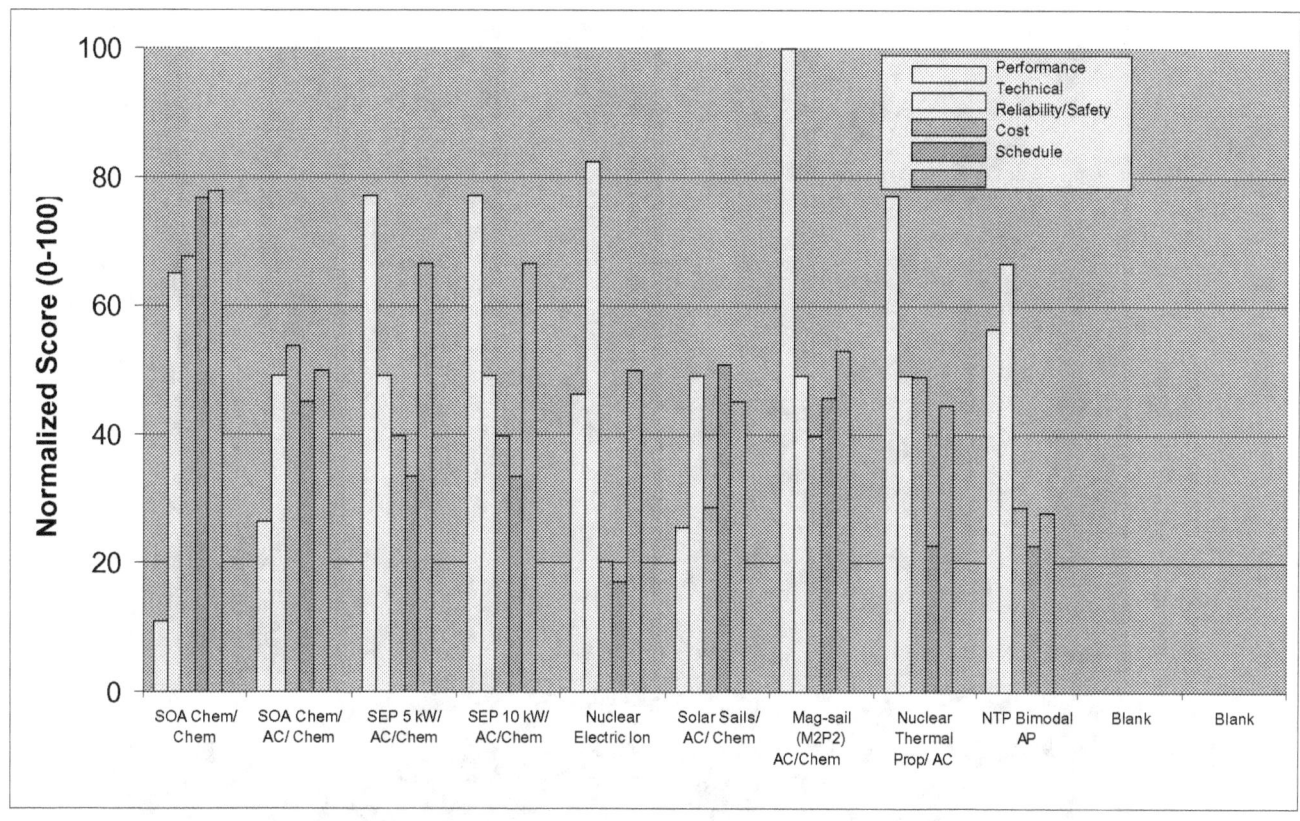

Figure C-5. Titan Explorer--Normalized Scores by Major Evaluation Area

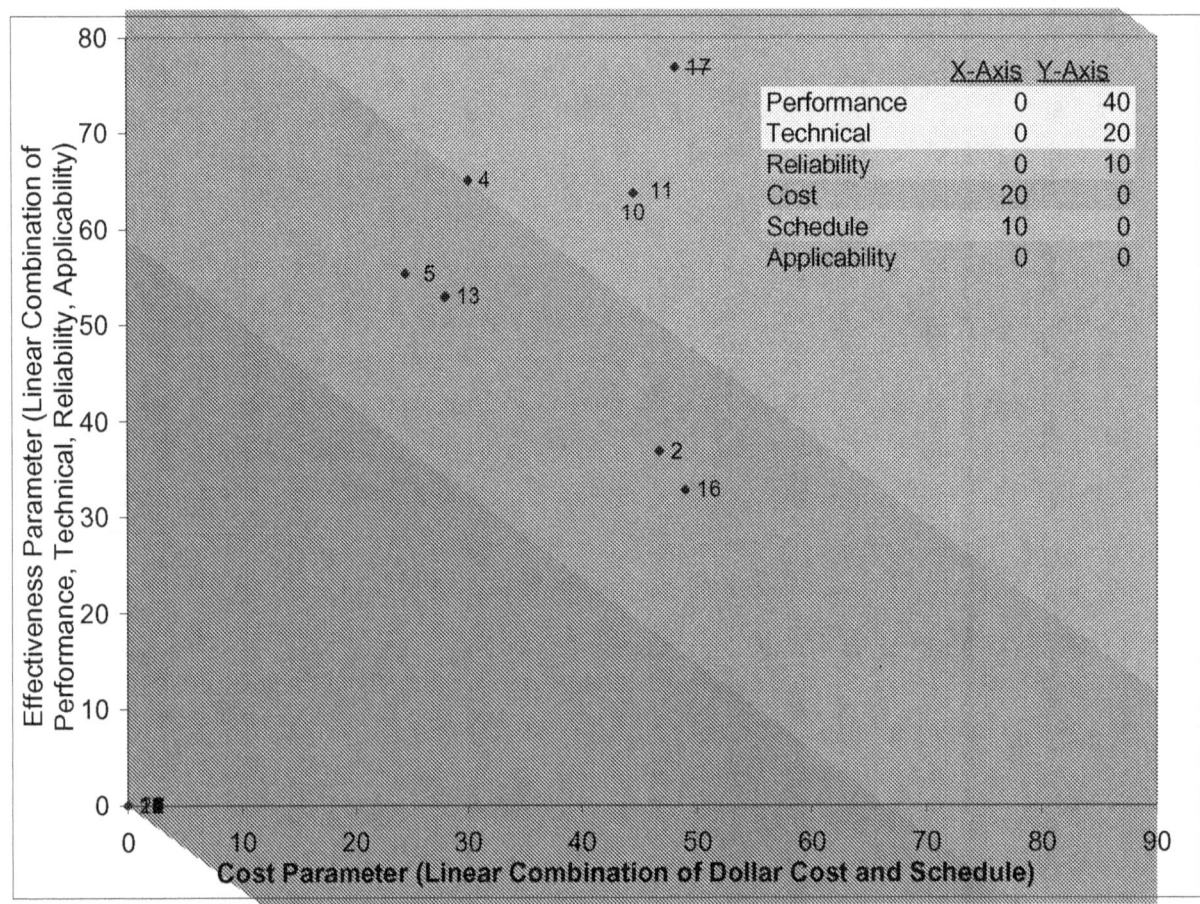

Figure C-6. Titan Explorer Cost-Effectiveness

2   SOA Chem/AC/Chem
4   Nuclear Thermal/AC
5   NTP Bimodal/AP
10 SEP(5 kW)/AC/Chem
11 SEP (10 kW)/AC/Chem
13 NEP Ion
16 Solar Sails/AC/Chem
17 Mag-sail (M2P2)/AC/Chem
AP = All Propulsion
AC = Aero Capture
SOA = State-of-the-art
MX = Momentum Exchange
NEP = Nuclear Electric

		FIGURES OF MERIT	WEIGHT	SOA Chm / AP	SOA Chm / AC	Adv Chm / AP	Nuclear Thermal Prop/AC	NTP Bi-modal AP	MX Tthr / Augmentation
		Technology Number		1	2	3	4	5	6
PERFORM.	1	Payload Mass Fraction	9	3	3	3	9	9	9
	2	Trip Time	10	9	9	9	9	9	9
	4	Time on Station	0						
	19	Prop. System Launch Mass & Volume	10	3	3	3	3	3	9
		Normalized Total	100	56.32	56.32	56.32	77.01	77.01	100.00
TECHNICAL	3	Operational Complexity	9	9	3	9	3	9	1
	5	Propellant Storage Time	9	9	9	3	9	3	9
	6	Station Keeping Precision	0						
	14	Crew Productivity	0						
	15	Sensitivity to Malfunctions	10	3	3	3	3	3	3
	16	Sensitivity to Perf. Deficiencies	10	9	3	9	3	9	3
	17	Enable In-Space Abort Scenarios	0						
	18	Crew/Payload Exposure to In-Space Env's	0						
		Normalized Total	100	82.46	49.12	66.67	49.12	66.67	43.86
RELIABILITY/ SAFETY	31	Pre-Launch Env. Hazards & Prot.	10	1	1	1	1	3	1
	41	In-Space Env. Hazards & Prot.	10	3	3	3	1	1	3
	42	Crew Exposure & Safety	0						
	43	Payload Exposure & Protection	6	3	9	9	9	3	9
	51	Relative Reliability Assessment	10	9	3	9	3	3	1
	53	Operating Life	10	9	9	9	9	3	9
		Normalized Total	100	56.48	53.70	67.59	49.07	28.70	49.07
COST	61	Technology Advancement Cost	9	9	9	3	9	1	1
	62	Mission Non-Recurring Cost	9	9	3	9	1	1	1
	67	Operational Cost	10	1	1	1	3	3	3
	68	Mission Recurring Cost	10	9	9	9	3	3	9
		Normalized Total	100	76.61	45.03	76.61	22.81	22.81	40.35
SCHEDULE	81	Total Development Time	10	9	3	3	3	3	3
	82	Special Facility Requirements	9	9	9	9	1	1	9
	83	Architectural Fragility	9	1	3	3	9	3	9
	84	Maturity (TRL Level)	8	9	3	3	3	3	1
		Normalized Total	100	77.78	50.00	50.00	44.44	27.78	55.56

Figure C-7. Europa Lander Weights and Scores (1 of 2)

	FIGURES OF MERIT	WEIGHT	ED Tthr / Augm-entation	Solr Elc /Chem: x kW	Nuclear Electric: Ion	Solar Sails / AC/Chm	Mag-sail (M2P2) / AP
	Technology Number		22	11	13	16	18
PERFORM.	1 Payload Mass Fraction	9		9	9	9	9
	2 Trip Time	10		9	9	3	9
	4 Time on Station	0					
	19 Prop. System Launch Mass & Volume	10		3	1	1	9
	Normalized Total	100	0.00	77.01	69.35	46.36	100.00
TECHNICAL	3 Operational Complexity	9		3	9	3	9
	5 Propellant Storage Time	9		9	9	9	9
	6 Station Keeping Precision	0					
	14 Crew Productivity	0					
	15 Sensitivity to Malfunctions	10		3	3	3	3
	16 Sensitivity to Perf. Deficiencies	10		3	9	3	9
	17 Enable In-Space Abort Scenarios	0					
	18 Crew/Payload Exposure to In-Space Env's	0					
	Normalized Total	100	0.00	49.12	82.46	49.12	82.46
RELIABILITY/ SAFETY	31 Pre-Launch Env. Hazards & Prot.	10	9	1	3	1	1
	41 In-Space Env. Hazards & Prot.	10		3	1	3	3
	42 Crew Exposure & Safety	0					
	43 Payload Exposure & Protection	8		9	1	3	3
	51 Relative Reliability Assessment	10		3	3	3	9
	53 Operating Life	10		3	1	3	3
	Normalized Total	100	20.83	39.81	20.37	28.70	42.59
COST	61 Technology Advancement Cost	9		3	1	3	1
	62 Mission Non-Recurring Cost	9		3	1	3	3
	67 Operational Cost	10		3	3	3	3
	68 Mission Recurring Cost	10		3	1	9	9
	Normalized Total	100	0.00	33.33	16.96	50.88	45.61
SCHEDULE	81 Total Development Time	10		3	3	3	1
	82 Special Facility Requirements	9		9	3	3	9
	83 Architectural Fragility	9		9	9	9	9
	84 Maturity (TRL Level)	8		3	3	1	0
	Normalized Total	100	0.00	66.67	50.00	45.06	53.09

Figure C-7. Europa Lander Weights and Scores (2 of 2)

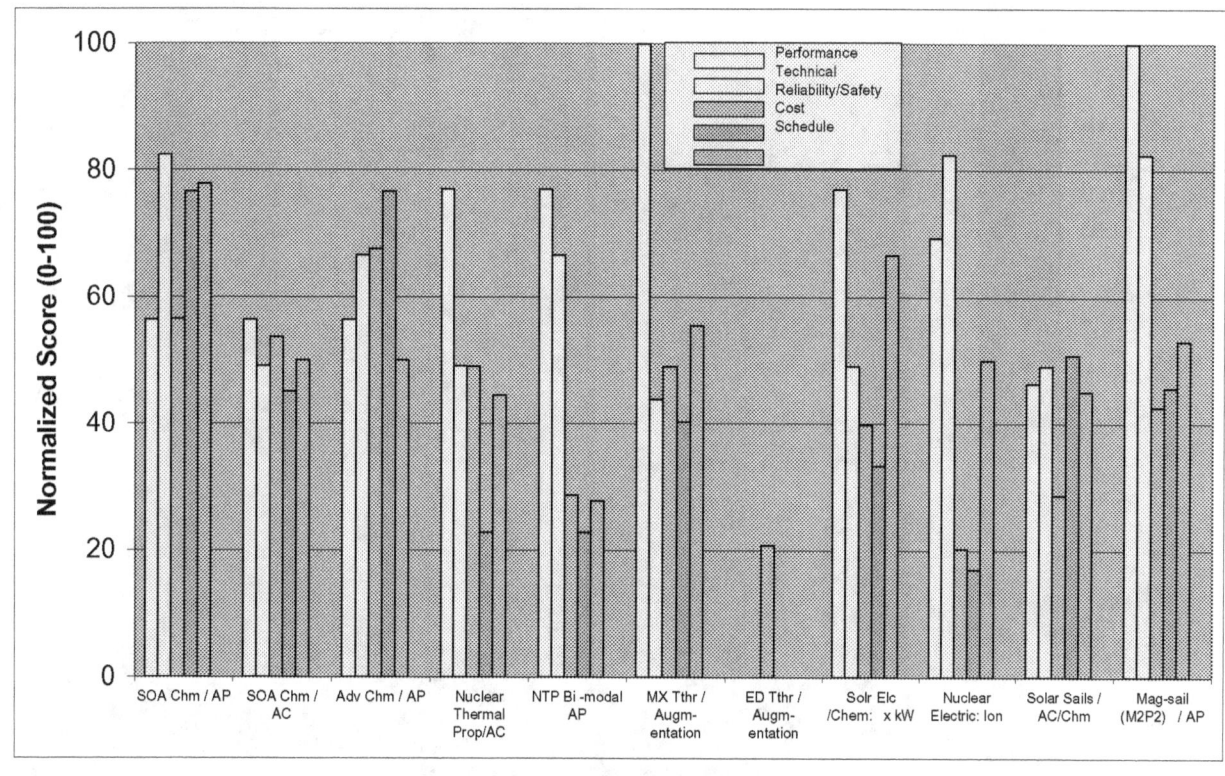

Figure C-8. Europa Lander--Normalized Score by Major Evaluation Area

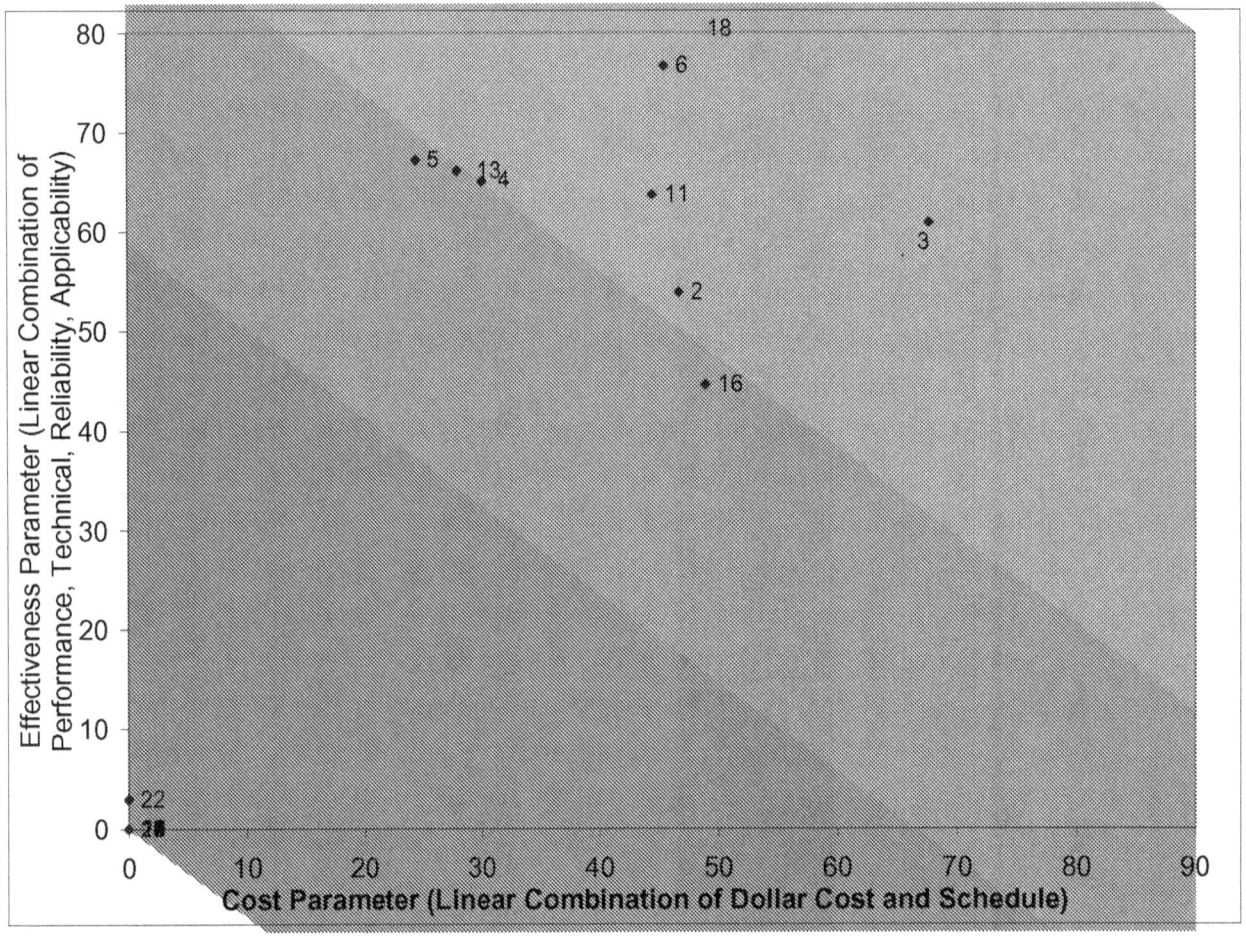

Figure C-9. Europa Lander Cost-Performance

2 SOA Chem/AC/Chem
3 Adv. Chem/Chem
4 Nuclear Thermal/AC
5 NTP Bimodal/AP
6 MX Tether/Augment./AP
11 SEP (10 kW)/AC/Chem
13 NEP Ion
16 Solar Sails/AC/Chem
18 Mag-sail (M2P2) AP

AP = All Propulsion
AC = Aero Capture
SOA = State-of-the-art
MX = Momentum Exchange
NEP = Nuclear Electric Propulsion
NTP = Nuclear Thermal Propulsion
SEP = Solar Electric Propulsion
MX = Momentum Exchange

		FIGURES OF MERIT	WEIGHT	SOA Chm / AP	SOA Chm / AC	Adv Chm / AP	Adv Chm / AC	Nuclear Thermal Prop	NTP Bi-modal
		Technology Number		1	2	3	24	4	5
PERFORM.	1	Payload Mass Fraction	9	3	9	3	9	9	9
	2	Trip Time	10	9	9	9	9	9	9
	4	Time on Station	0						
	19	Prop. System Launch Mass & Volume	10	3	9	3	9	9	9
		Normalized Total	100	56.32	100.00	56.32	100.00	100.00	100.00
TECHNICAL	3	Operational Complexity	9	9	3	9	3	3	9
	5	Propellant Storage Time	9	9	9	3	3	9	3
	6	Station Keeping Precision	0						
	14	Crew Productivity	0						
	15	Sensitivity to Malfunctions	10	3	3	3	3	3	3
	16	Sensitivity to Perf. Deficiencies	10	9	3	9	3	3	9
	17	Enable In-Space Abort Scenarios	0						
	18	Crew/Payload Exposure to In-Space Env's	0						
		Normalized Total	100	82.46	49.12	66.67	33.33	49.12	66.67
RELIABILITY/ SAFETY	31	Pre-Launch Env. Hazards & Prot.	10	1	1	1	1	1	1
	41	In-Space Env. Hazards & Prot.	10	3	3	3	3	1	3
	42	Crew Exposure & Safety	0						
	43	Payload Exposure & Protection	10	3	9	9	9	9	9
	51	Relative Reliability Assessment	10	9	3	9	3	3	1
	53	Operating Life	10	9	9	9	9	9	9
		Normalized Total	100	55.56	55.56	68.89	55.56	51.11	51.11
COST	61	Technology Advancement Cost	9	9	9	3	9	3	1
	62	Mission Non-Recurring Cost	9	9	3	9	3	1	1
	67	Operational Cost	10	1	3	1	3	3	3
	68	Mission Recurring Cost	10	9	9	9	9	3	3
		Normalized Total	100	76.61	50.88	76.61	50.88	22.81	22.81
SCHEDULE	81	Total Development Time	10	9	3	3	3	3	3
	82	Special Facility Requirements	9	9	9	9	9	1	1
	83	Architectural Fragility	9	3	3	3	3	9	9
	84	Maturity (TRL Level)	8	9	3	3	3	3	3
		Normalized Total	100	83.33	50.00	50.00	50.00	44.44	44.44

Figure C-10. Mars Sample Return Weights and Scores (1 of 2)

	FIGURES OF MERIT	WEIGHT	SEP w/ sep. lander	NEP	Mag-sail AC	Mag-sail AP
	Technology Number		25	13	17	18
PERFORM.	1 Payload Mass Fraction	9	3	9	9	9
	2 Trip Time	10	9	9	9	9
	4 Time on Station	0				
	19 Prop. System Launch Mass & Volume	10	9	9	9	9
	Normalized Total	100	79.31	100.00	100.00	100.00
TECHNICAL	3 Operational Complexity	9	3	9	3	9
	5 Propellant Storage Time	9	9	9	9	9
	6 Station Keeping Precision	0				
	14 Crew Productivity	0				
	15 Sensitivity to Malfunctions	10	3	3	3	3
	16 Sensitivity to Perf. Deficiencies	10	9	9	3	9
	17 Enable In-Space Abort Scenarios	0				
	18 Crew/Payload Exposure to In-Space Env's	0				
	Normalized Total	100	66.67	82.46	49.12	82.46
RELIABILITY/ SAFETY	31 Pre-Launch Env. Hazards & Prot.	10	1	1	1	1
	41 In-Space Env. Hazards & Prot.	10	3	1	3	3
	42 Crew Exposure & Safety	0				
	43 Payload Exposure & Protection	10	9	1	9	3
	51 Relative Reliability Assessment	10	3	3	3	9
	53 Operating Life	10	3	3	9	9
	Normalized Total	100	42.22	20.00	55.56	55.56
COST	61 Technology Advancement Cost	9	3	1	1	1
	62 Mission Non-Recurring Cost	9	3	1	3	3
	67 Operational Cost	10	3	3	3	3
	68 Mission Recurring Cost	10	9	1	9	9
	Normalized Total	100	50.88	16.96	45.61	45.61
SCHEDULE	81 Total Development Time	10	3	3	1	1
	82 Special Facility Requirements	9	9	3	9	9
	83 Architectural Fragility	9	9	9	9	9
	84 Maturity (TRL Level)	8	3	3	0	0
	Normalized Total	100	66.67	50.00	53.09	53.09

Figure C-10. Mars Sample Return Weights and Scores (2 of 2)

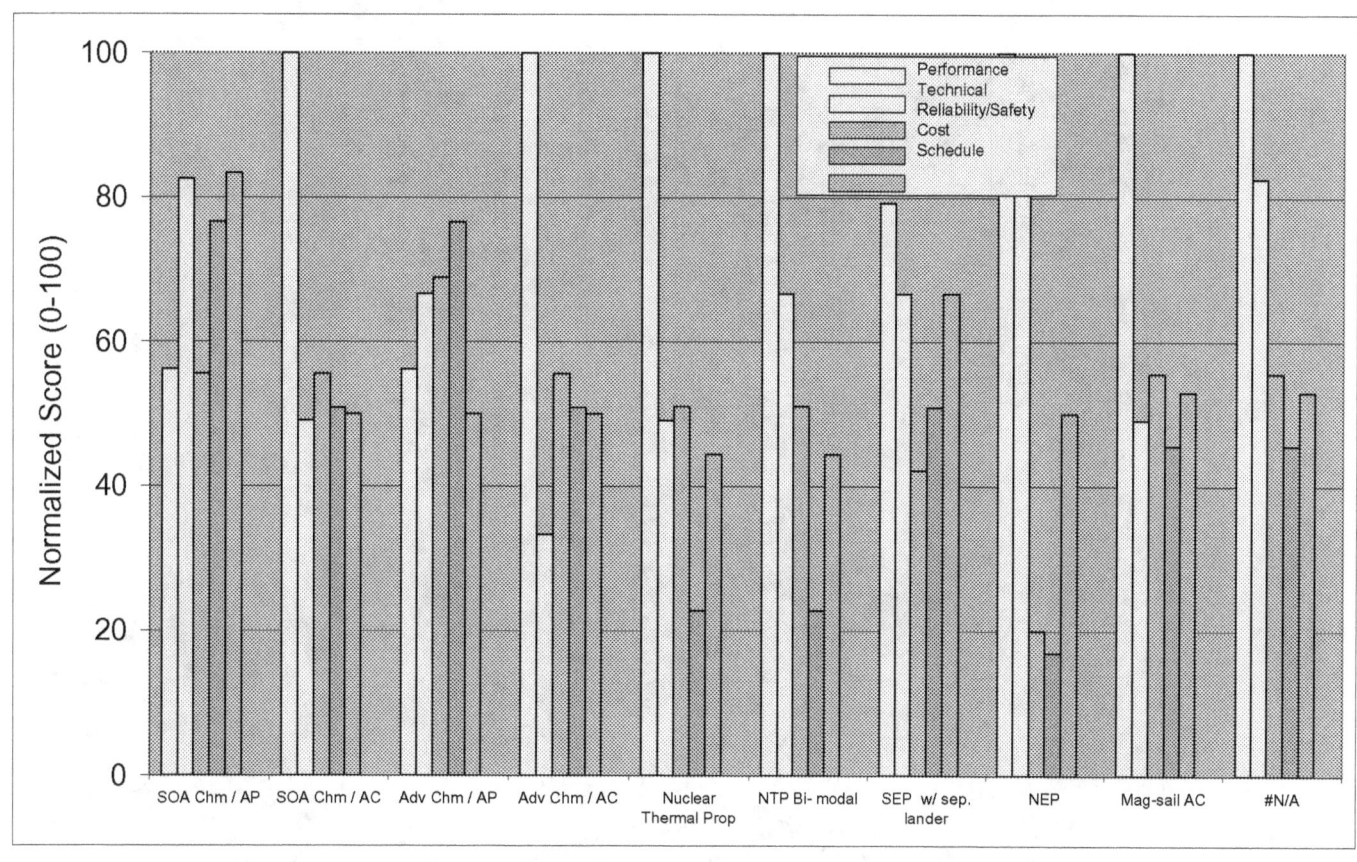

Figure C-11. Mars Sample Return--Normalized Score by Major Evaluation Area

IISTP Phase I Final Report
September 14, 2001

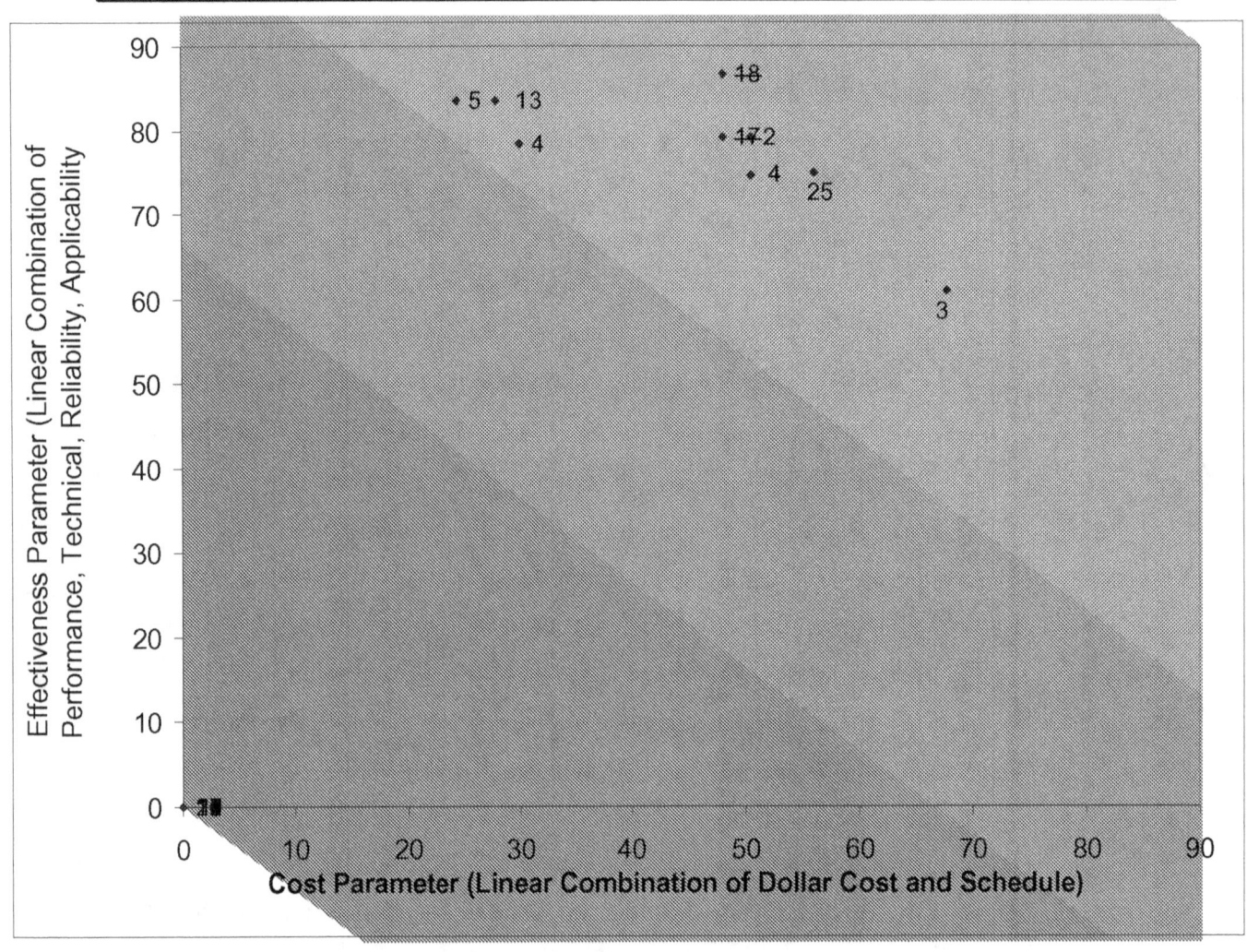

Figure C-12. Mars Sample Return Cost-Performance Scatter Chart

   2 SOA Chem/AC/Chem
   3 Adv. Chem/Chem
   4 Nuclear Thermal/AC
   5 NTP Bimodal/AP
 13 NEP Ion
 17 Mag-sail (M2P2)/AC/Chem
 18 Mag-sail (M2P2) AP
 24 Adv. Chem/AC
 25 SEP w/ separate lander
AP = All Propulsion
AC = Aero Capture
SOA = State-of-the-art
MX = Momentum Exchange
NEP = Nuclear Electric Propulsion
NTP = Nuclear Thermal Propulsion
SEP = Solar Electric Propulsion
MX = Momentum Exchange

	FIGURES OF MERIT	WEIGHT	Solar Sails	Nucl-ear Elec	NEP	NTP / NEP Hybrid	Mag-sail M2P2
	Technology Number		16	12	13	20	17
PERFORM.	1 Payload Mass Fraction	9	9	1	1	1	9
	2 Trip Time	10	3	3	3	3	9
	4 Time on Station	0					
	19 Prop. System Launch Mass & Volume	10	3	3	3	3	3
	Normalized Total	100	54.02	26.44	26.44	26.44	77.01
TECHNICAL	3 Operational Complexity	9	9	3	3	3	9
	5 Propellant Storage Time	9	9	9	9	9	1
	6 Station Keeping Precision	0					
	14 Crew Productivity	0					
	15 Sensitivity to Malfunctions	10	3	3	3	3	3
	16 Sensitivity to Perf. Deficiencies	10	3	9	9	9	9
	17 Enable In-Space Abort Scenarios	0					
	18 Crew/Payload Exposure to In-Space Env's	0					
	Normalized Total	100	64.91	66.67	66.67	66.67	61.40
RELIABILITY/ SAFETY	31 Pre-Launch Env. Hazards & Prot.	10	1	3	3	3	1
	41 In-Space Env. Hazards & Prot.	10	9	3	9	1	9
	42 Crew Exposure & Safety	0					
	43 Payload Exposure & Protection	8	9	3	9	3	9
	51 Relative Reliability Assessment	10	3	3	3	1	3
	53 Operating Life	10	3	1	1	1	3
	Normalized Total	100	53.70	28.70	53.70	19.44	53.70
COST	61 Technology Advancement Cost	9	1	1	1	1	1
	62 Mission Non-Recurring Cost	9	3	1	1	1	3
	67 Operational Cost	10	3	1	1	3	3
	68 Mission Recurring Cost	10	3	1	1	1	9
	Normalized Total	100	28.07	11.11	11.11	16.96	45.61
SCHEDULE	81 Total Development Time	10	1	1	1	3	1
	82 Special Facility Requirements	9	9	3	3	1	9
	83 Architectural Fragility	9	1	3	3	3	3
	84 Maturity (TRL Level)	8	1	1	1	3	1
	Normalized Total	100	33.33	22.22	19.75	27.78	38.89

Figure C-13. Interstellar Probe Weights and Scores

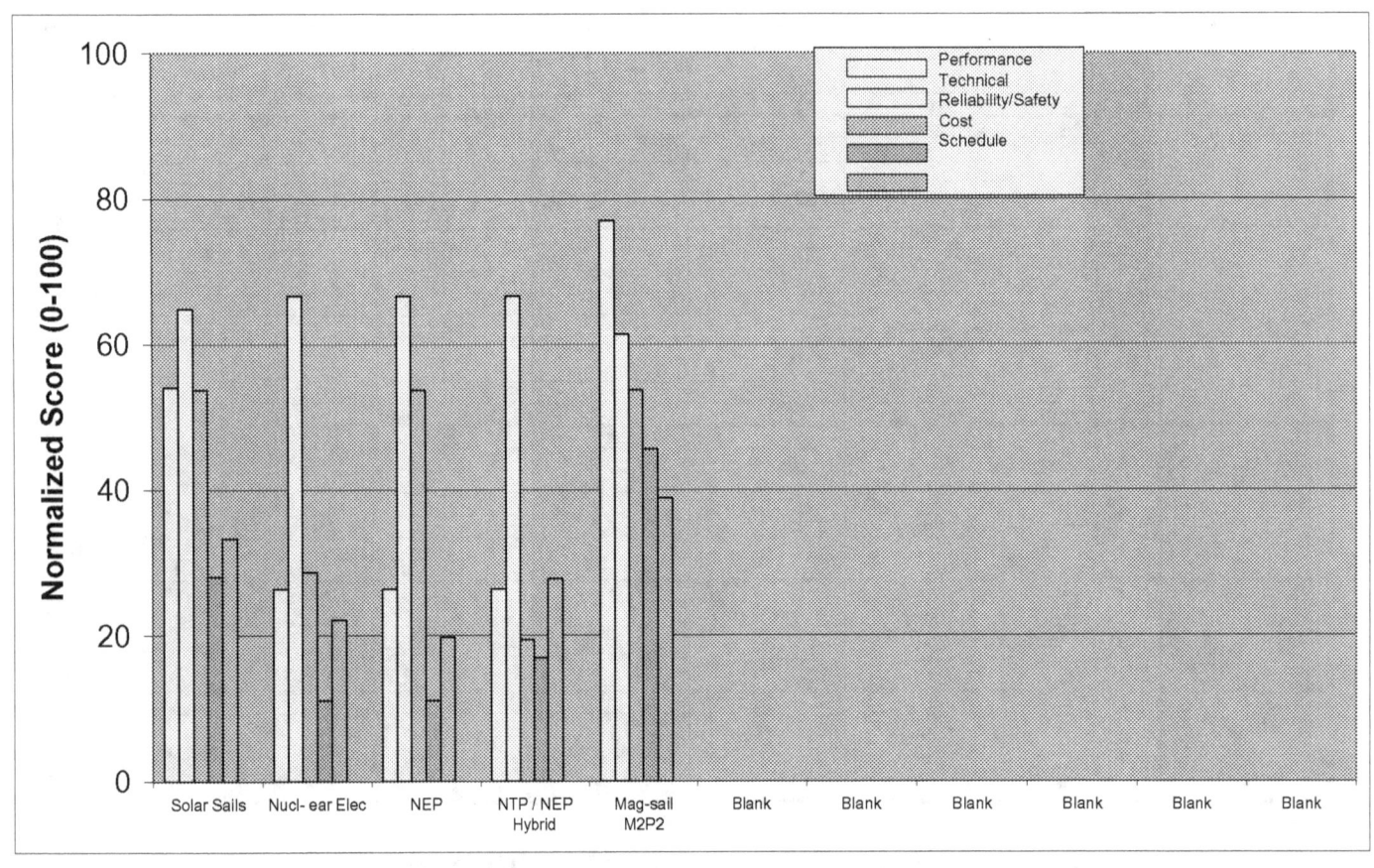

Figure C-14. Interstellar Probe--Normalized Score by Major Evaluation Area

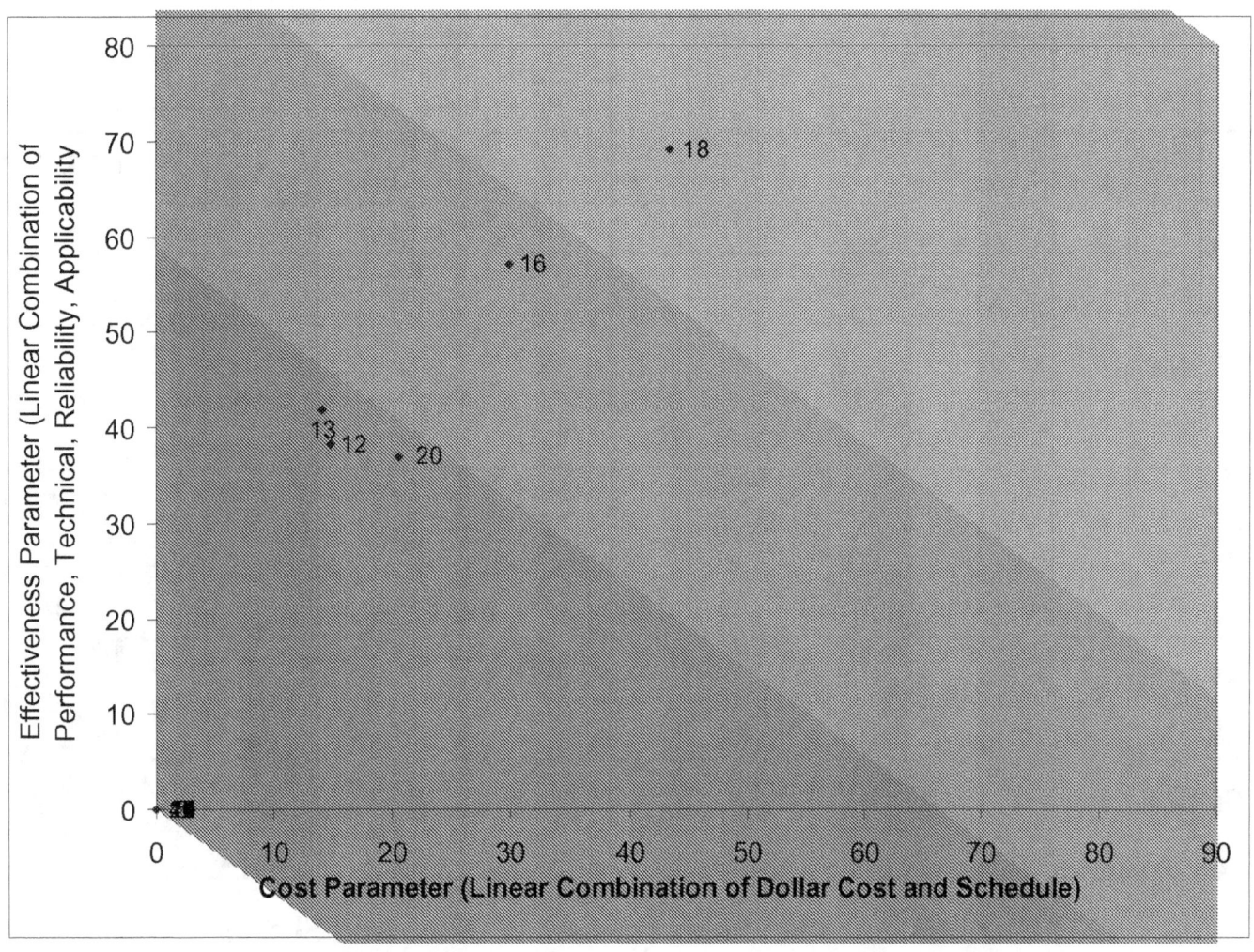

Figure C-15. Interstellar Probe Cost-Performance Scatter Chart

12 NEP Hall/Chem/AC/Chem
13 NEP Ion
16 Solar Sails/AC/Chem
17 Mag-sail (M2P2)/AC/Chem
18 Mag-sail (M2P2) AP
20 NTP/NEP Hybrid
AP = All Propulsion
AC = Aero Capture
SOA = State-of-the-art
MX = Momentum Exchange
NEP = Nuclear Electric Propulsion
NTP = Nuclear Thermal Propulsion
SEP = Solar Electric Propulsion
MX = Momentum Exchange

		FIGURES OF MERIT	WEIGHT	Solar Sails	Solar Electric	SEP/ Chem SEP	Nuclear Elec
		Technology Number		16	11	10	13
PERFORM.	1	Payload Mass Fraction	9	9	3	3	1
	2	Trip Time	10	3	9	3	3
	4	Time on Station	0				
	19	Prop. System Launch Mass & Volume	10	3	9	3	1
		Normalized Total	100	54.02	79.31	33.33	18.77
TECHNICAL	3	Operational Complexity	9	9	9	3	3
	5	Propellant Storage Time	9				
	6	Station Keeping Precision	0	9	9	9	9
	14	Crew Productivity	0				
	15	Sensitivity to Malfunctions	10	9	3	3	3
	16	Sensitivity to Perf. Deficiencies	10	9	9	9	9
	17	Enable In-Space Abort Scenarios	0				
	18	Crew/Payload Exposure to In-Space Env's	0				
		Normalized Total	100	76.32	58.77	42.98	42.98
RELIABILITY/ SAFETY	31	Pre-Launch Env. Hazards & Prot.	10	9	9	3	3
	41	In-Space Env. Hazards & Prot.	10	9	9	3	9
	42	Crew Exposure & Safety	0				
	43	Payload Exposure & Protection	8	9	9	9	9
	51	Relative Reliability Assessment	10	9	9	9	9
	53	Operating Life	10	9	9	9	3
		Normalized Total	100	100.00	100.00	72.22	72.22
COST	61	Technology Advancement Cost	9				
	62	Mission Non-Recurring Cost	9				
	67	Operational Cost	10				
	68	Mission Recurring Cost	10				
		Normalized Total	100	0.00	0.00	0.00	0.00
SCHEDULE	81	Total Development Time	10	3	3	3	3
	82	Special Facility Requirements	9	9	9	9	3
	83	Architectural Fragility	9	9	9	9	9
	84	Maturity (TRL Level)	8	3	3	3	3
		Normalized Total	100	66.67	66.67	66.67	50.00

Figure C-16. Solar Polar Imager Weights and Scores

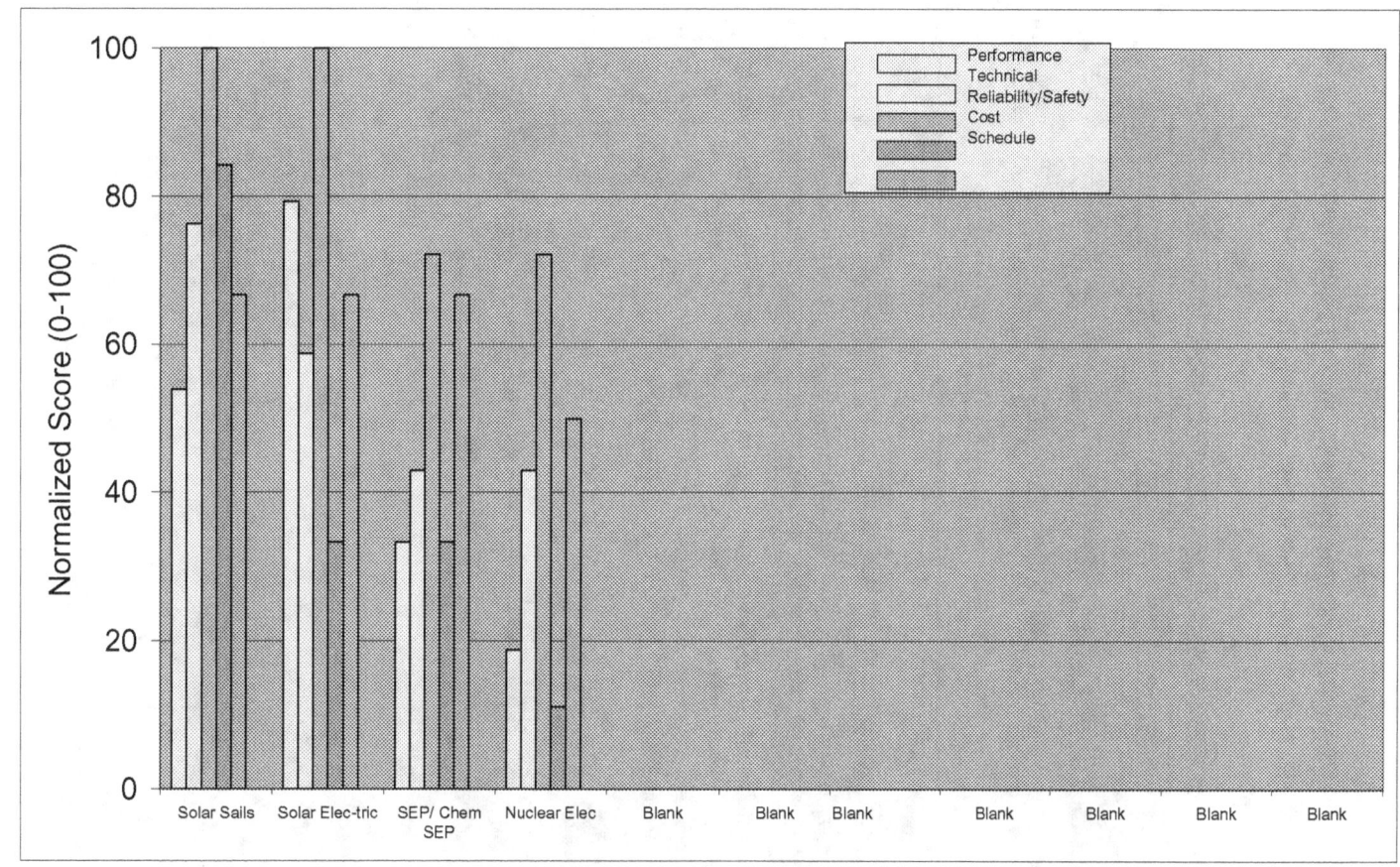

Figure C-17. Solar Polar Imager--Normalized Score by Major Evaluation Area

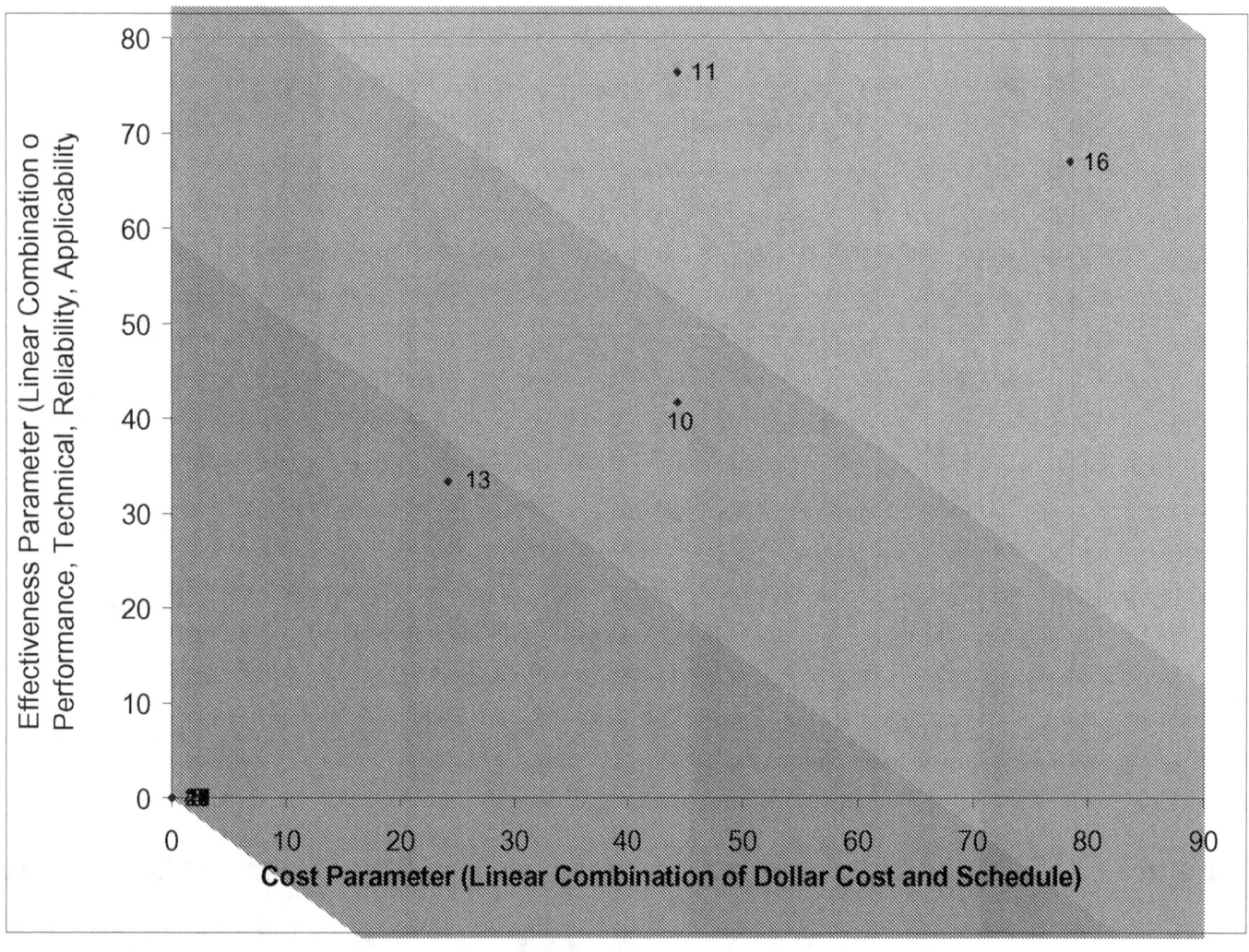

Figure C-18. Solar Polar Imager Cost-Performance Scatter Chart

10 SEP(5 kW)/AC/Chem
11 SEP (10 kW)/AC/Chem
13 NEP Ion
16 Solar Sails/AC/Chem
AP = All Propulsion
AC = Aero Capture
SOA = State-of-the-art
MX = Momentum Exchange
NEP = Nuclear Electric Propulsion
NTP = Nuclear Thermal Propulsion
SEP = Solar Electric Propulsion
MX = Momentum Exchange

		FIGURES OF MERIT	WEIGHT	SOA Chem/ Chem	Adv. Chem	Solar Thermal Prop.	Solar Sail	Solar Elec: 5kW Hall	Solar Elec: 10kW Ion
				1	3	21	16	10	11
PERFORM.	1	Payload Mass Fraction	10	1	1	3	1	9	9
	2	Trip Time	5	9	9	9	3	3	3
	4	Time on Station	5						
	19	Prop. System Launch Mass & Volume	0	1	3	3	1	9	3
		Normalized Total	100	30.56	30.56	41.67	13.89	58.33	58.33
TECHNICAL	3	Operational Complexity	5	3	3	3	1	1	1
	5	Propellant Storage Time	5	9	3	1	9	9	9
	6	Station Keeping Precision	2						
	14	Crew Productivity	0						
	15	Sensitivity to Malfunctions	8	3	3	3	3	3	3
	16	Sensitivity to Perf. Deficiencies	7	9	9	9	9	9	9
	17	Enable In-Space Abort Scenarios	2						
	18	Crew/Payload Exposure to In-Space Env's	5						
		Normalized Total	100	48.04	38.24	34.97	44.77	44.77	44.77
RELIABILITY/ SAFETY	31	Pre-Launch Env. Hazards & Prot.	5	3	9	9	9	9	9
	41	In-Space Env. Hazards & Prot.	8	3	9	9	9	9	9
	42	Crew Exposure & Safety	0						
	43	Payload Exposure & Protection	8	9	9	9	9	3	3
	51	Relative Reliability Assessment	8	9	9	9	9	9	9
	53	Operating Life	7	9	9	9	9	9	9
		Normalized Total	100	75.93	100.00	100.00	100.00	85.19	85.19
COST	61	Technology Advancement Cost	8	9	3	3	3	9	3
	62	Mission Non-Recurring Cost	8	9	3	3	3	3	3
	67	Operational Cost	9	1	1	3	1	9	9
	68	Mission Recurring Cost	9	9	9	9	3	3	3
		Normalized Total	100	76.47	45.10	50.98	27.45	66.67	50.98
SCHEDULE	81	Total Development Time	5	9	3	3	3	9	3
	82	Special Facility Requirements	5	9	9	9	9	9	9
	83	Architectural Fragility	5	1	3	3	9	9	9
	84	Maturity (TRL Level)	5	9	3	3	3	9	3
		Normalized Total	100	77.78	50.00	50.00	66.67	100.00	66.67

Figure C-19. Magnetospheric Constellation Weights and Scores

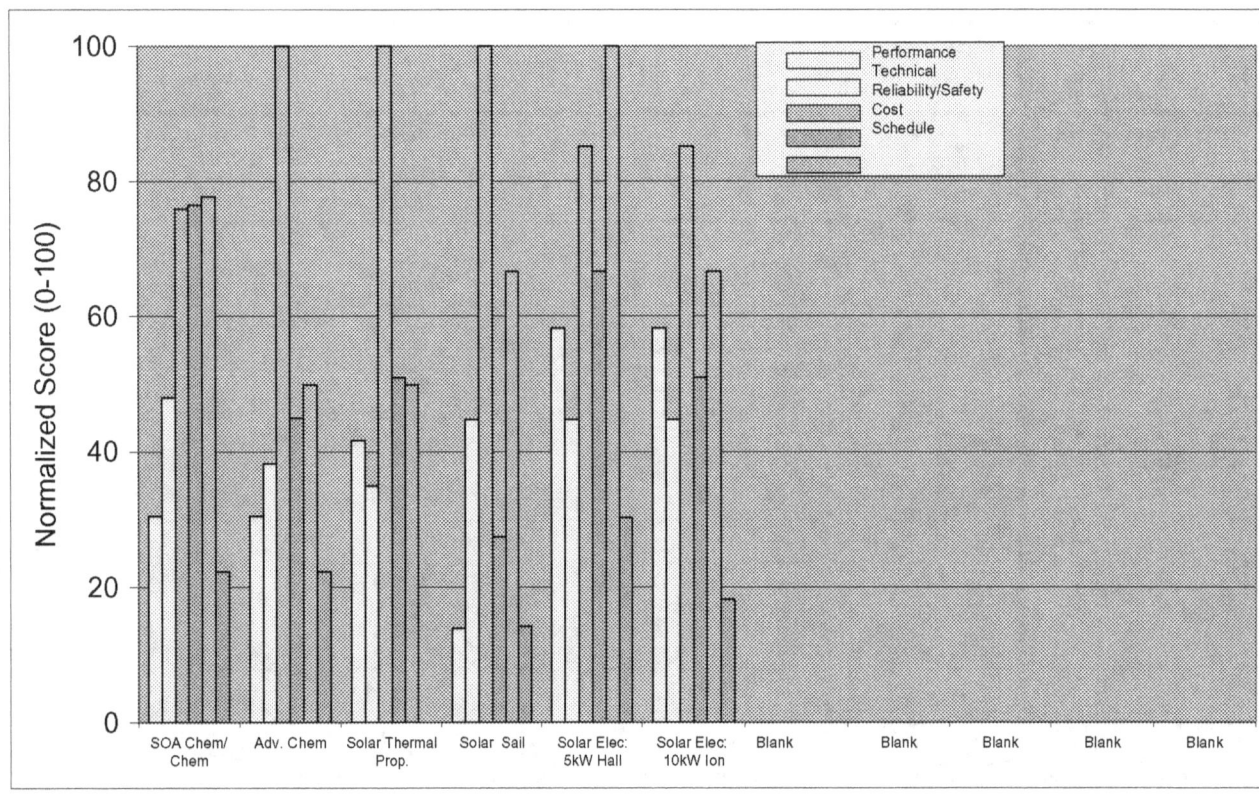

Figure C-20. Magnetospheric Constellation Normalized Scores by Major Evaluation Area

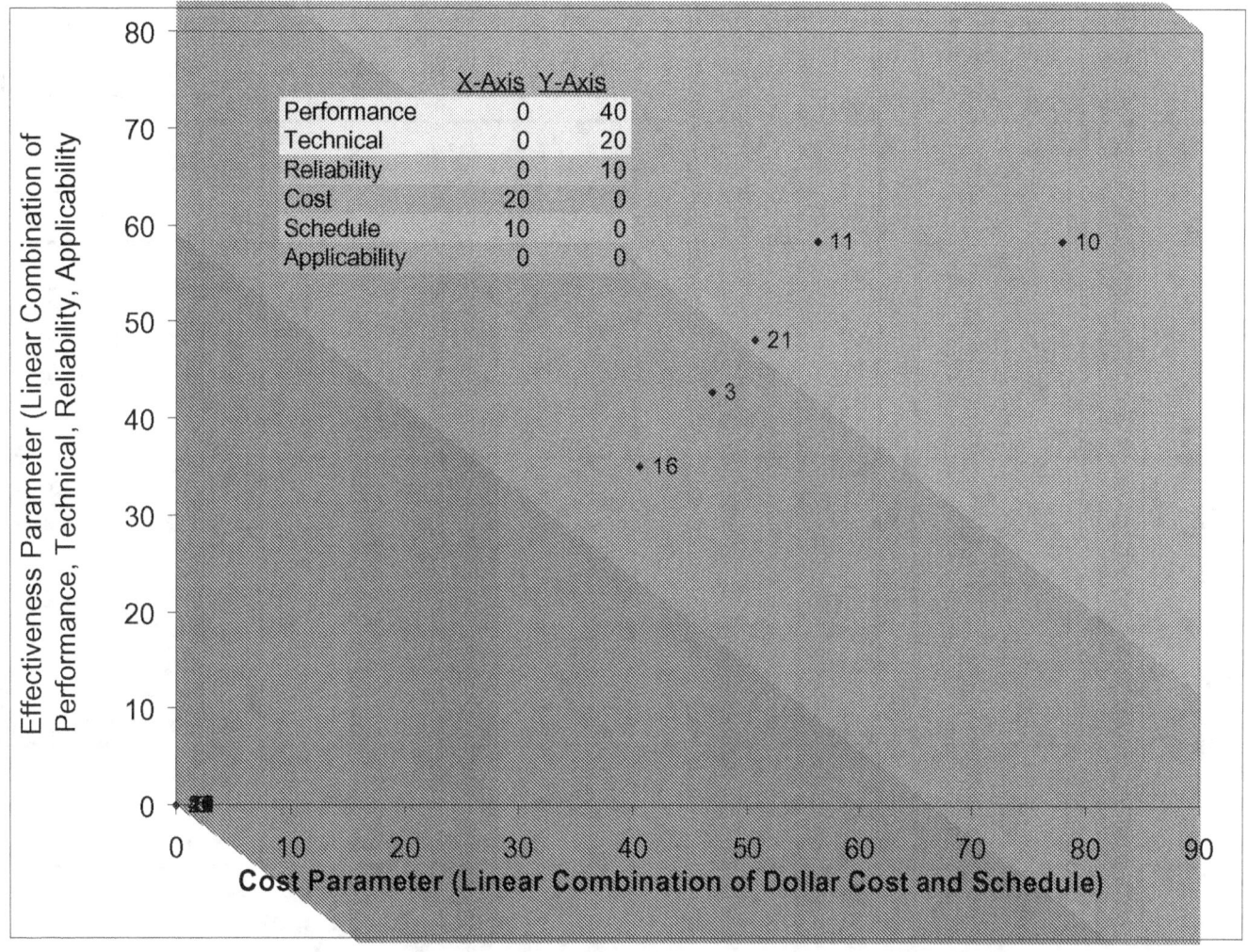

Figure C-21. Magnetospheric Constellation Cost-Performance Scatter Chart

3 Adv. Chem.
10 SEP(5 kW)
11 SEP (10 kW)
16 Solar Sails
21 Solar Thermal Prop
AP = All Propulsion
AC = Aero Capture
SOA = State-of-the-art
MX = Momentum Exchange
NEP = Nuclear Electric Propulsion
NTP = Nuclear Thermal Propulsion
SEP = Solar Electric Propulsion
MX = Momentum Exchange

		FIGURES OF MERIT	WEIGHT	Solar Sails	Solar Elec NSTAR	Solar Elec 5 kW Ion	Solar Elc 10 kW Ion	Solar Elc 5 kW Hall
				16	23	10	11	9
PERFORM.	1	Payload Mass Fraction	10	3	9	9	9	9
	2	Trip Time	5	9	3	3	3	3
	4	Time on Station	5	9	3	3	3	3
	19	Prop. System Launch Mass & Volume	0	3	3	3	3	3
		Normalized Total	100	66.67	66.67	66.67	66.67	66.67
TECHNICAL	3	Operational Complexity	5	9	3	3	3	3
	5	Propellant Storage Time	5	9	9	9	9	9
	6	Station Keeping Precision	2	9	9	9	9	9
	14	Crew Productivity	0					
	15	Sensitivity to Malfunctions	8	3	3	3	3	3
	16	Sensitivity to Perf. Deficiencies	7	9	9	9	9	9
	17	Enable In-Space Abort Scenarios	2					
	18	Crew/Payload Exposure to In-Space Env's	5					
		Normalized Total	100	63.73	53.92	53.92	53.92	53.92
RELIABILITY/ SAFETY	31	Pre-Launch Env. Hazards & Prot.	5	9	9	9	9	9
	41	In-Space Env. Hazards & Prot.	8	9	3	3	3	3
	42	Crew Exposure & Safety	0					
	43	Payload Exposure & Protection	8	9	3	3	3	3
	51	Relative Reliability Assessment	8	3	9	9	9	9
	53	Operating Life	7	9	9	9	9	3
		Normalized Total	100	85.19	70.37	70.37	70.37	57.41
COST	61	Technology Advancement Cost	8	3	9	3	3	9
	62	Mission Non-Recurring Cost	8	3	9	9	9	9
	67	Operational Cost	9	3	9	9	9	9
	68	Mission Recurring Cost	9	3	3	3	3	9
		Normalized Total	100	33.33	82.35	50.98	50.98	100.00
SCHEDULE	81	Total Development Time	5	3	9	3	3	9
	82	Special Facility Requirements	5	9	9	9	9	9
	83	Architectural Fragility	5	9	9	9	9	9
	84	Maturity (TRL Level)	5	3	9	3	3	9
		Normalized Total	100	66.67	100.00	66.67	66.67	100.00

Figure C-22. Pole Sitter Weights and Scores

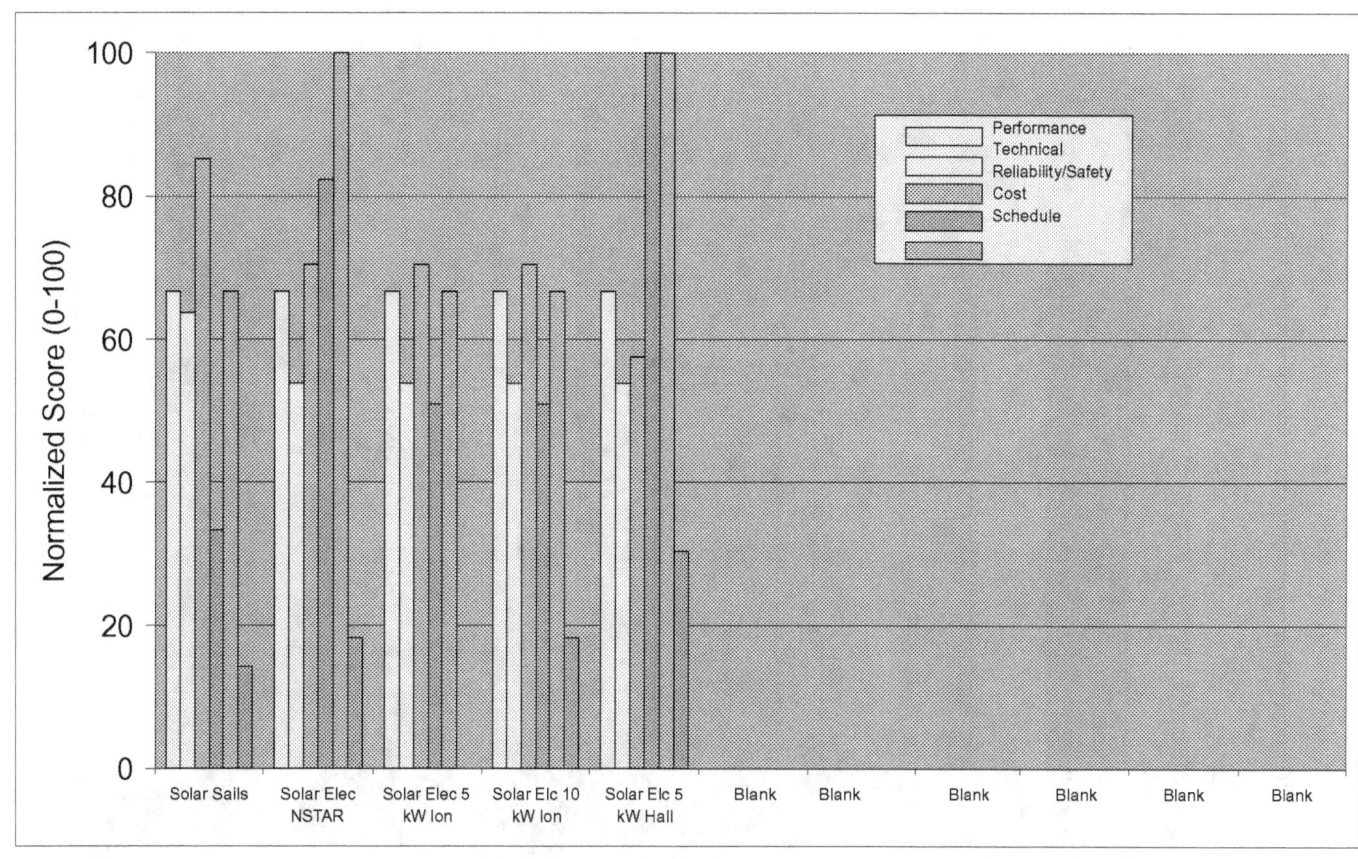

Figure C-23. Pole Sitter Normalized Scores by Major Evaluation Area

IISTP Phase I Final Report
September 14, 2001

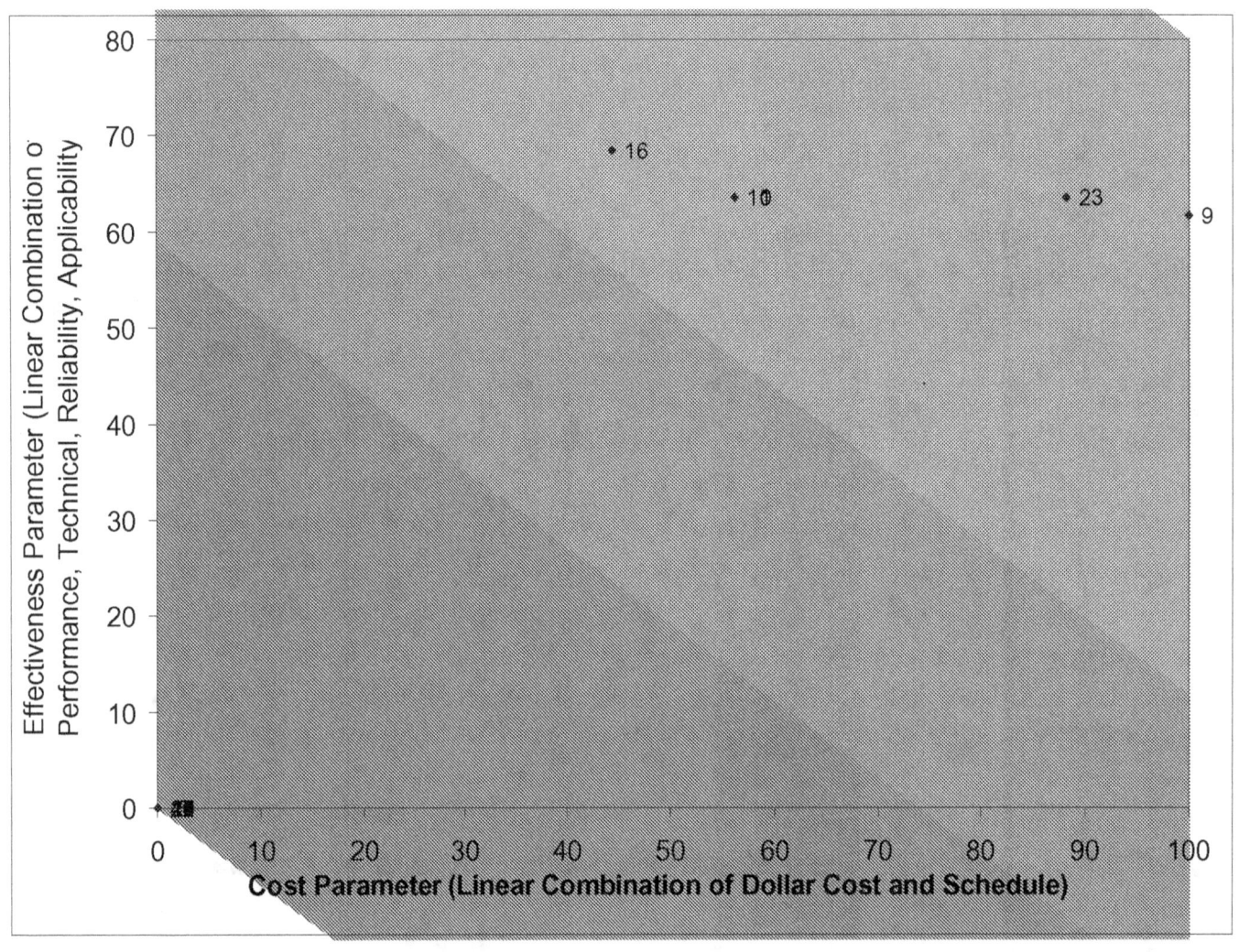

Figure C-24. Pole Sitter Cost-Performance Scatter Chart

9 SEP Hall/Chem/AC/Chm
10 SEP(5 kW)/AC/Chem
11 SEP (10 kW)/AC/Chem
16 Solar Sails/AC/Chem
23 SEP NSTAR
AP = All Propulsion
AC = Aero Capture
SEP = Solar Electric Propulsion

		FIGURES OF MERIT	WEIGHT	Nuclear Thermal Prop/AC	NTP Bimodal AP	Solar Elec 100 kW AC/Chem	Nuclear Elec Ion/MPD	Nuclear Elec VaSIMR	NTP/NEP Hubrid
				4	5	9	13	14	20
PERFORM.	1	Payload Mass Fraction	5	3	3	3	3	3	9
	2	Trip Time	10	9	9	9	9	9	9
	4	Time on Station	0						
	19	Prop. System Launch Mass & Volume	1	3	3	9	3	3	3
		Normalized Total	100	75.00	75.00	79.17	75.00	75.00	95.83
TECHNICAL	3	Operational Complexity	5	3	9	1	1	1	1
	5	Propellant Storage Time	5	1	1	3	9	3	1
	6	Station Keeping Precision	0						
	14	Crew Productivity	5	9	9	9	3	3	3
	15	Sensitivity to Malfunctions	5	1	3	1	9	9	3
	16	Sensitivity to Perf. Deficiencies	5	1	3	3	9	9	9
	17	Enable In-Space Abort Scenarios	7	1	3	1	9	9	3
	18	Crew/Payload Exposure to In-Space Env's	3	9	9	9	9	9	9
		Normalized Total	100	34.60	54.92	34.60	77.78	68.25	42.22
RELIABILITY/ SAFETY	31	Pre-Launch Env. Hazards & Prot.	0	3	3	9	3	3	3
	41	In-Space Env. Hazards & Prot.	2	3	3	9	3	3	3
	42	Crew Exposure & Safety	8	1	3	3	3	3	3
	43	Payload Exposure & Protection	8	9	9	3	3	3	9
	51	Relative Reliability Assessment	10	3	3	3	3	3	3
	53	Operating Life	0						
		Normalized Total	100	46.03	52.38	38.10	33.33	33.33	52.38
COST	61	Technology Advancement Cost	2	3	1	9	1	1	1
	62	Mission Non-Recurring Cost	10	3	3	9	3	3	1
	67	Operational Cost	7	9	9	3	1	1	3
	68	Mission Recurring Cost	7	3	3	9	3	3	1
		Normalized Total	100	51.28	49.57	82.05	25.64	25.64	17.09
SCHEDULE	81	Total Development Time	10	3	1	9	1	1	1
	82	Special Facility Requirements	3	1	1	9	3	3	1
	83	Architectural Fragility	5	3	3	3	9	9	3
	84	Maturity (TRL Level)	10	3	1	9	1	1	1
		Normalized Total	100	30.95	15.08	88.10	29.37	29.37	15.08

Figure C-25. Mars Piloted Weights and Scores

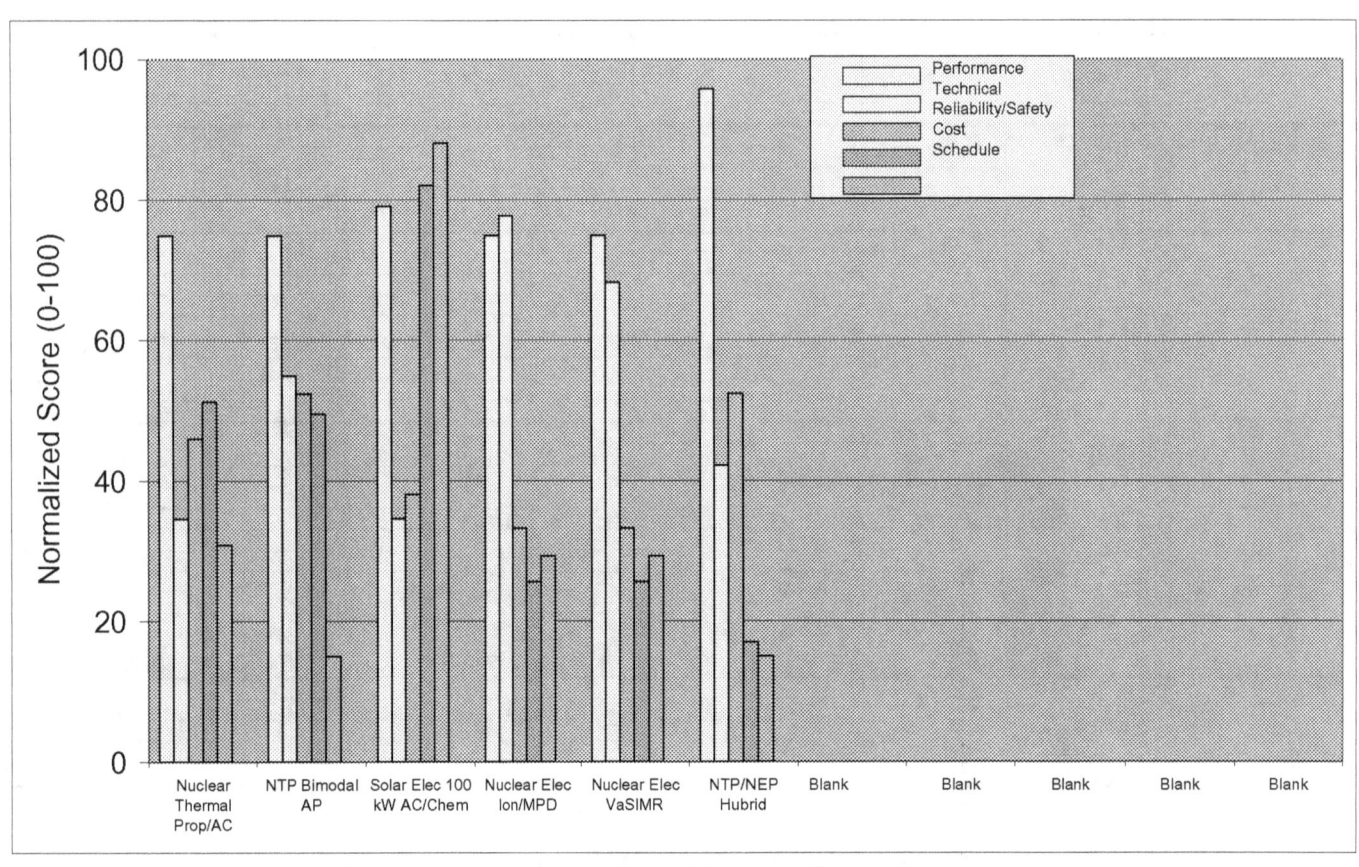

Figure C-26. Mars Piloted-- Normalized Scores by Major Evaluation Area

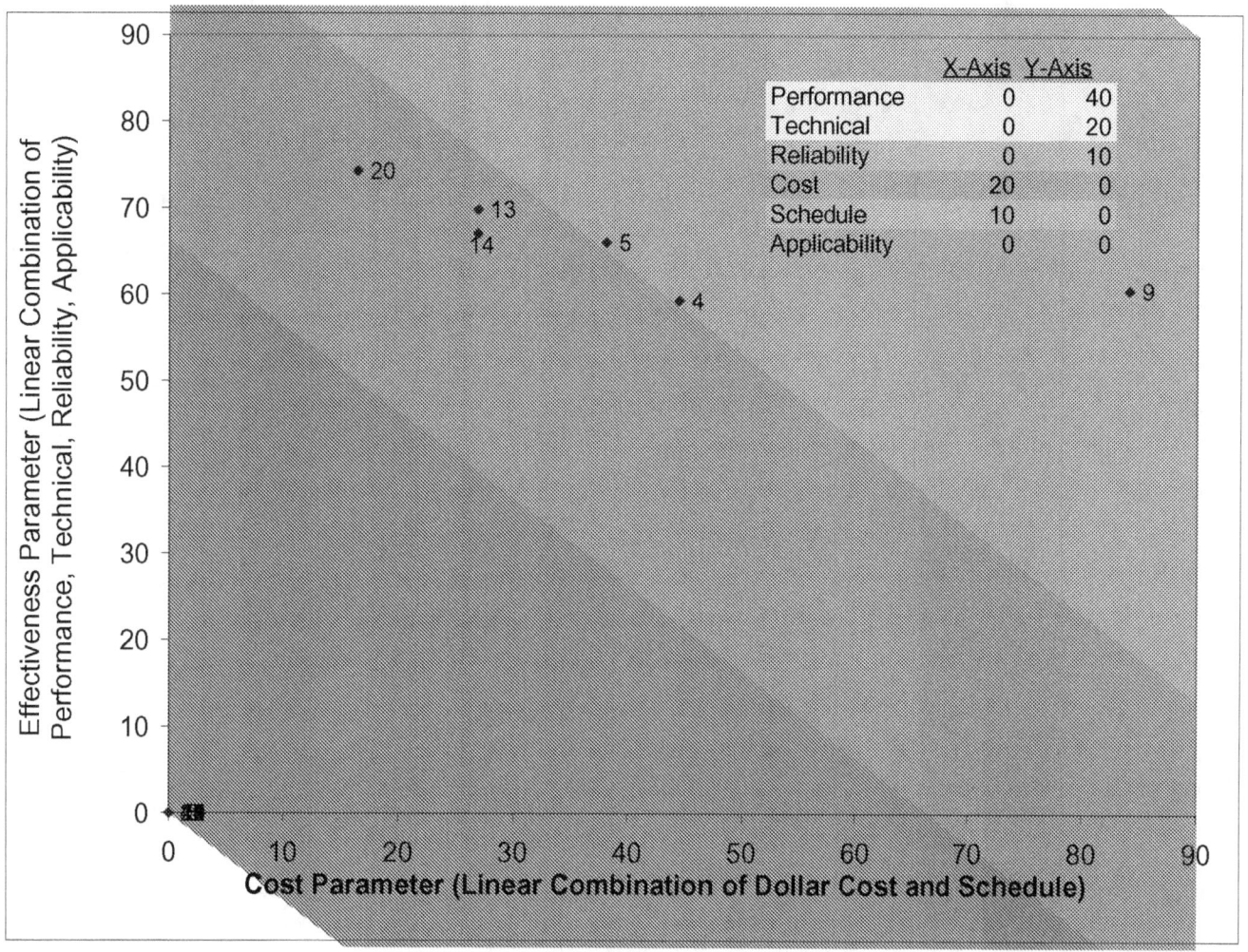

Figure C-27. Mars-Piloted Cost-Performance Scatter Chart

    4   Nuclear Thermal/AC
    5   NTP Bimodal/AP
   11 SEP (100 kW)/AC/Chem
   13 NEP Ion/MPD
   14 Nuclear Elec: VaSIMR
   20 NTP/NEP Hybrid
AP = All Propulsion
AC = Aero Capture
NEP = Nuclear Electric Propulsion
NTP = Nuclear Thermal Propulsion
SEP = Solar Electric Propulsion

IISTP Phase I Final Report
September 14, 2001

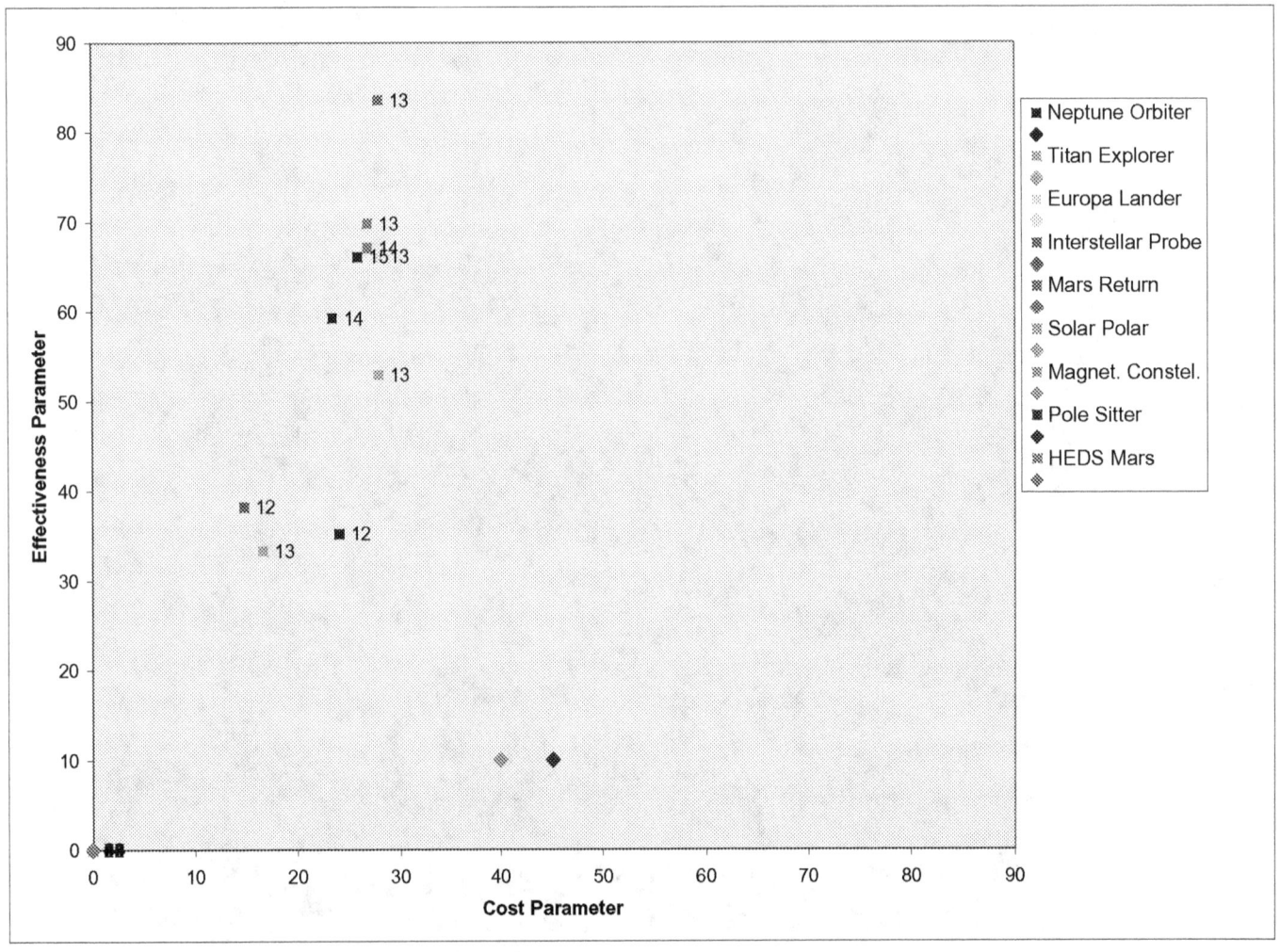

Figure C-28. Nuclear Electric Propulsion Across All Missions

12 NEP Hall/Chem/AC/Chem
13 NEP Ion
14 NEP VaSIMR
15 NEP MPD

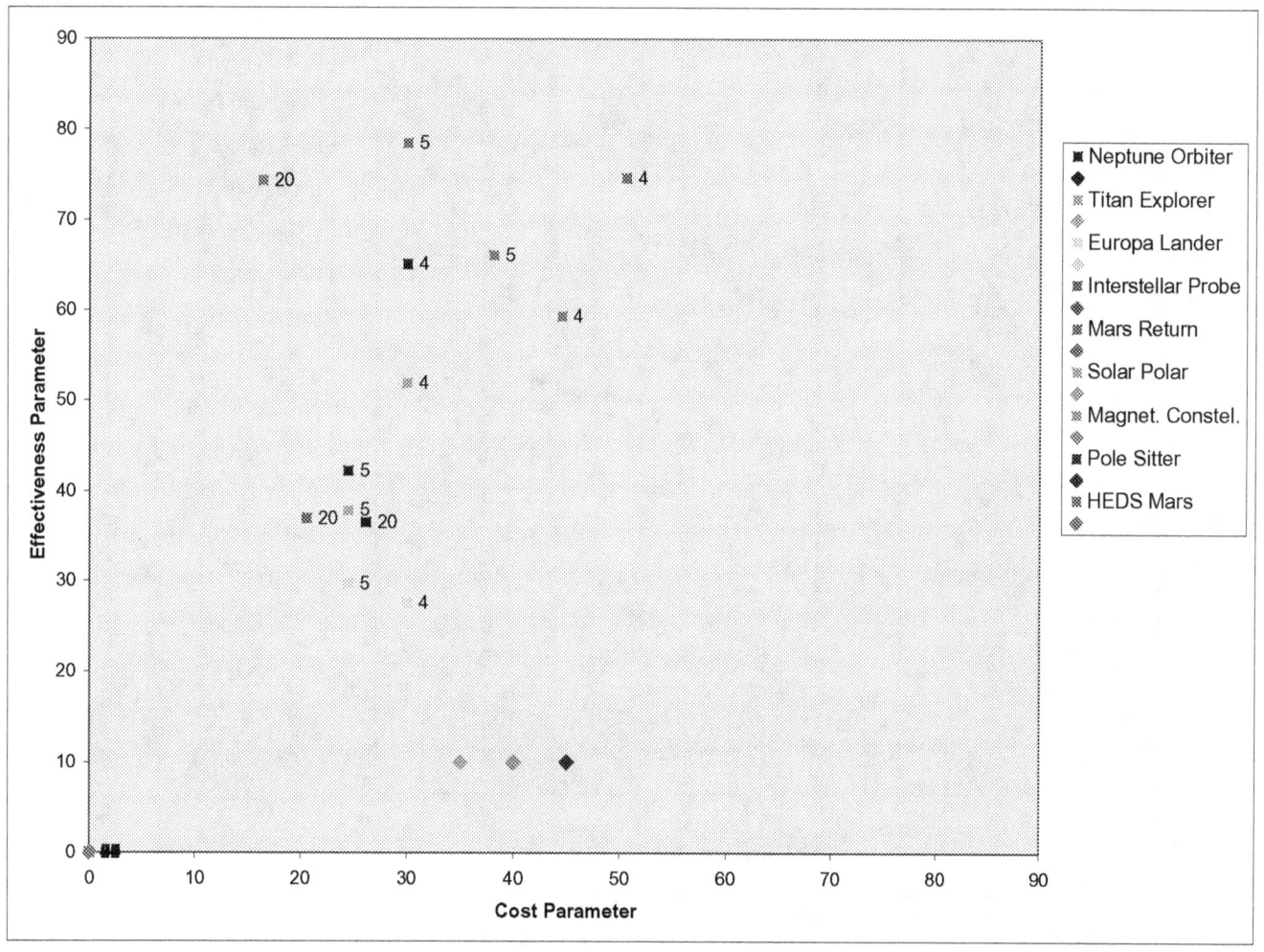

Figure C-29. Nuclear Thermal Propulsion Across All Missions

4 Nuclear Thermal/AC
5 NTP Bimodal/AP
20 NTP/NEP Ion Hybrid

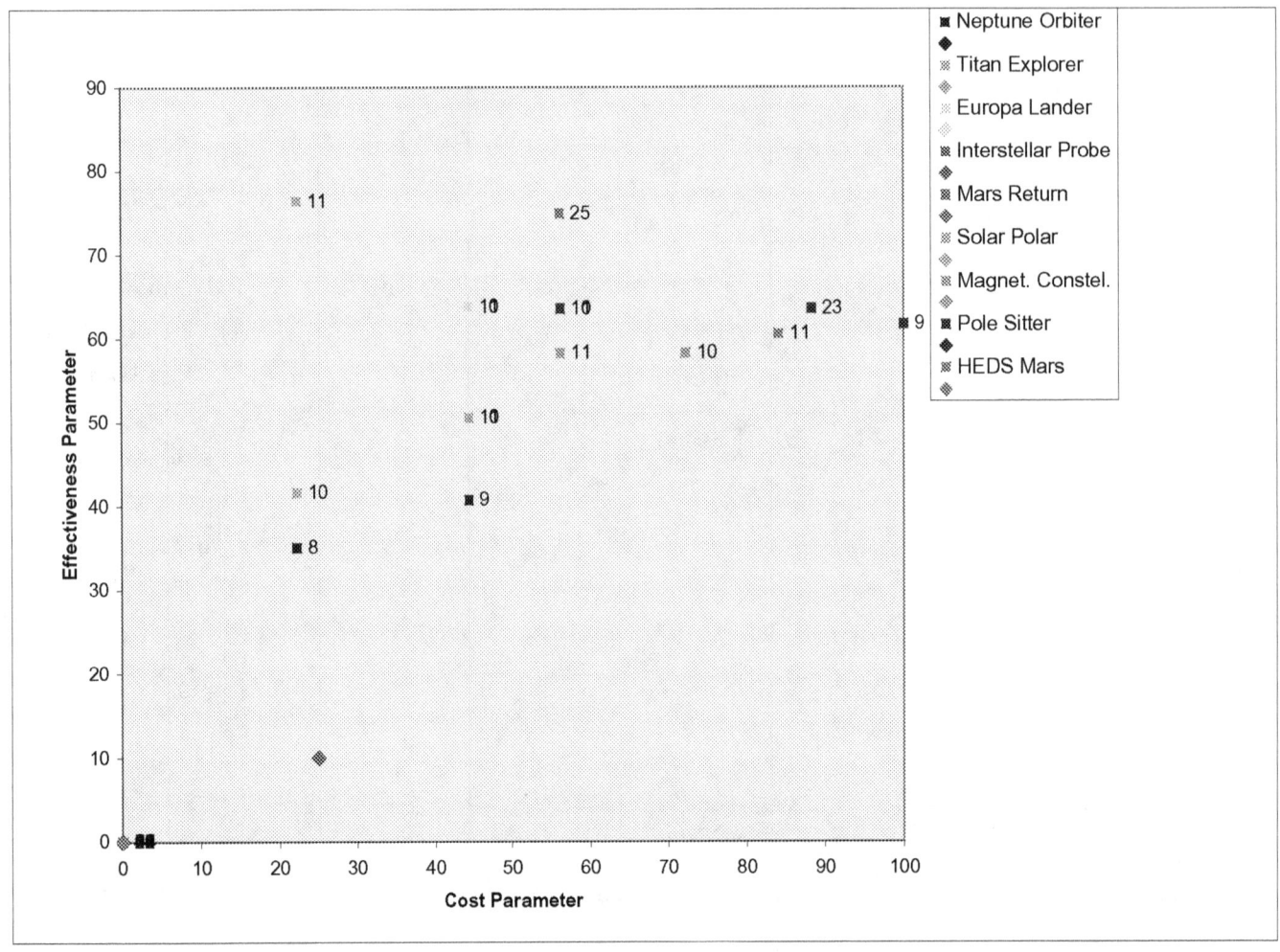

Figure C-30. Solar Electric Propulsion Across All Missions

8   SEP Hall/NTP/AC/Chem
9   SEP Hall/Chem/AC/Chm
10 SEP(5 kW) Ion /AC/Chem
11 SEP (10 kW) Ion /AC/Chem
23 SEP NSTAR
25 SEP Ion w/separate lander

IISTP Phase I Final Report
September 14, 2001

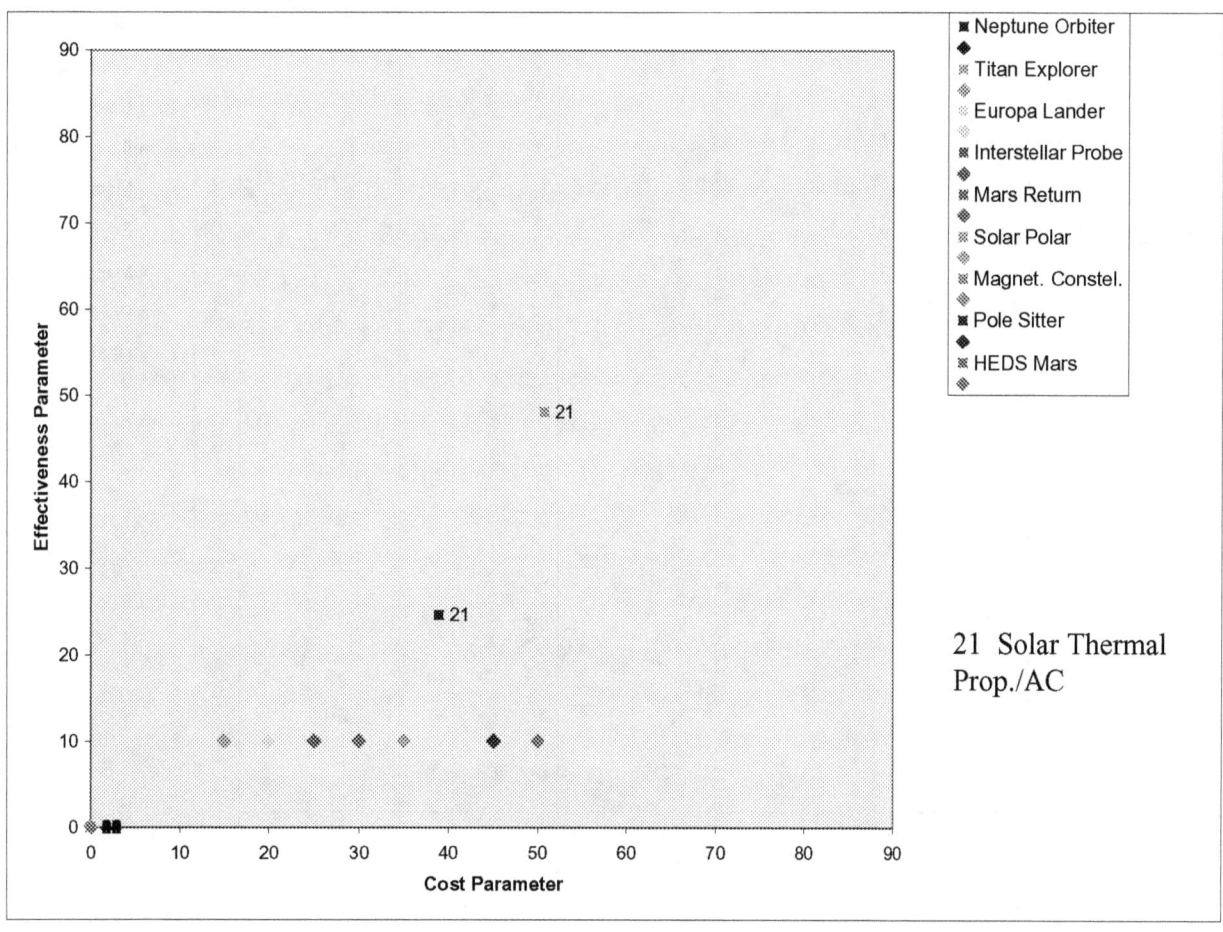

Figure C-31. Solar Thermal Propulsion Across All Missions

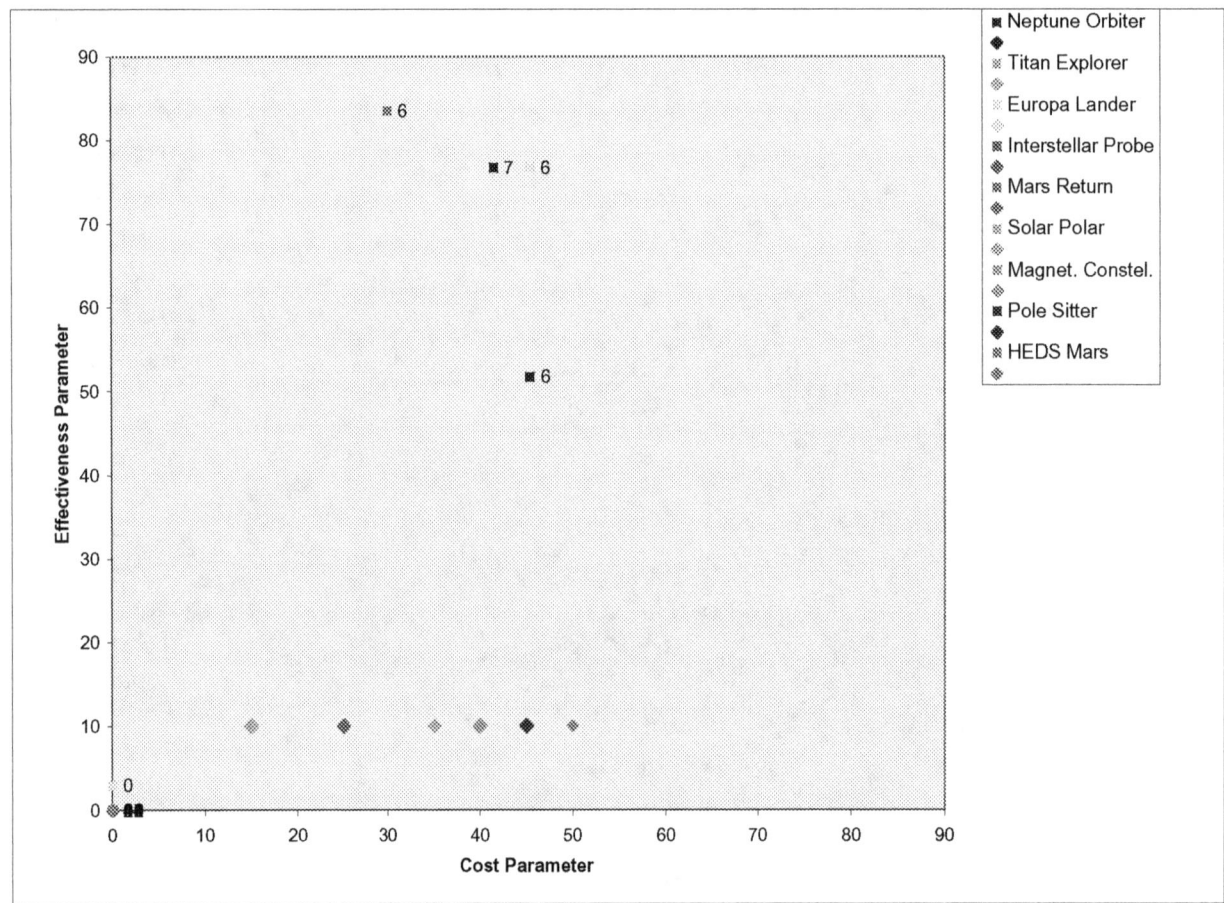

Figure C-32. Tether Augmentation Across All Missions

6 MX Tether/Augment./AP
7 MX Tether/Augment./AC/Chem
22 ED Tether Augmentation

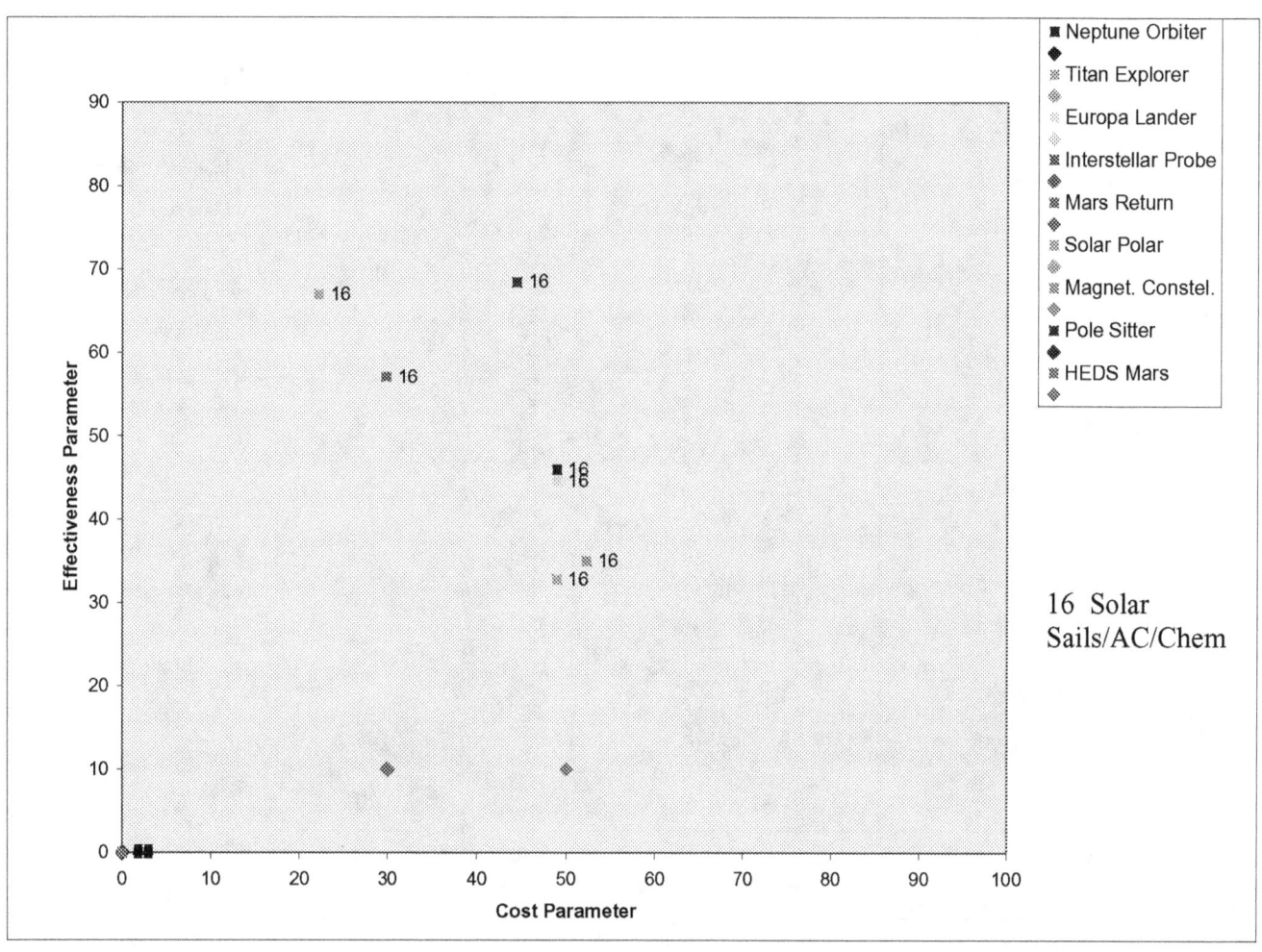

Figure C-33. Solar Sails Across All Missions

IISTP Phase I Final Report
September 14, 2001

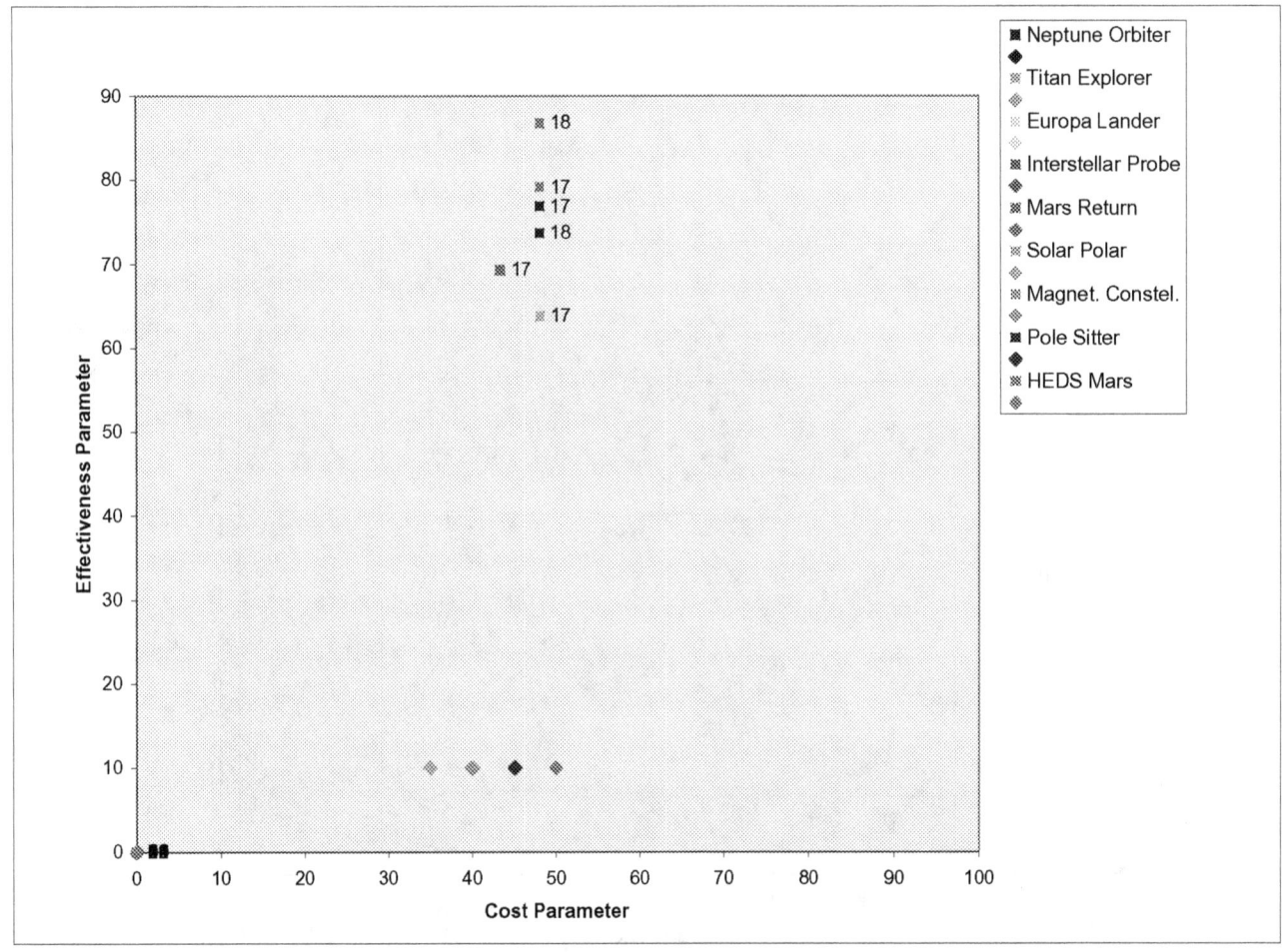

Figure C-34. Plasma Sails Across All Missions

17 Mag-sail (M2P2)/AC/Chem
18 Mag-sail (M2P2) AP

APPENDIX D

Mission Analyses Results

Mission analyses were performed by the Systems Team as needed to characterize the IISTP propulsion technologies' capabilities, as applied to each of the missions. This appendix consists of briefing charts prepared to report analyses of the Neptune Orbiter and Titan Explorer missions. Neptune Orbiter analyses were performed by Systems Team members at the various NASA centers. The Titan Explorer analyses were performed by JPL's Team X. Neptune Orbiter analyses considered all of the candidate technologies, and Titan Explorer analyses considered most of them. These results are representative of the IISTP mission analyses and provide insight into the performance capabilities of candidate propulsion technologies.

IISTP Phase I Final Report
September 14, 2001

In-space Integrated Space Transportation Activity

by the

IISTP Systems Analysis Team
Neptune Orbiter Mission Package

Friday, May 4th, 2001
Larry Kos

Current Team: MSFC, GRC, JPL, JSC, LaRC, Gray Research, SRS Technologies, Tethers Unlimited

In-space ISTP Activity
Systems Team (ST) Members

1.	Joe Bonometti	MFSC Systems, Solar Thermal
2.	Bob Cataldo	GRC Power Systems
3.	Bret Drake	JSC Systems, Human Missions
4.	Len Dudzinski	GRC Systems, NTP/NEP, Trajectories/Sizing, Fusion
5.	Bob Farris	Gray Research/TAC
6.	Robert Frisbee	JPL Systems, Sails
7.	Leon Gefert	GRC/Lead, Systems POC
8.	Jeff George	JSC/Lead, Systems, NEP
9.	Rob Hoyt	TU Tethers, Sizing
10.	Jonathan Jones	MSFC Plasma Sails, Technology Team Lead
11.	Larry Kos	MSFC/Lead, Chemical Trajectories/Sizing
12.	Melissa McGuire	GRC NTP Systems, Trajectories/Sizing
13.	Jim Moore	SRS Systems, ED Tethers, STP
14.	Michelle Munk	LaRC/Lead, Aeroassist
15.	Mahmoud Naderi	MSFC Cost
16.	Muriel Noca	JPL/Lead, Systems, & Team-X POC, Sizing
17.	Tara Poston	MSFC Trajectories/Sizing
18.	Bob Sefcik	GRC Cost
19.	Kirk Sorensen	MSFC MX Tethers
20.	Noble Stone	SRS Systems, ED Tethers, STP
21.	Gordon Woodcock	Gray Research/TAC
22.	Scott Baird	JSC ISPP Systems
23.	John Blandino	JPL POC for Code S
24.	Neil Dennehy	GSFC POC
25.	Sandy Kirkindall	MFSC Systems
26.	Saroj Patel	MFSC Systems
	Consultants for ST:	Juan Ayon, Chen-Wan Yen (JPL Sail & EP Trajectories)
		Steve Oleson (GRC SEP data)
		Steve Tucker, Dave Plachta (MSFC & GRC CFM)

In-space ISTP Activity
Proposed Missions (28) and Mission Categories (8)

#	Category	#	Mission
1	Earth vicinity, low to moderate ΔV	1	Geospace Electrodynamic Connections (GEC)
		2	EREMF (Leonardo)
		3	Hal SAR
		4	LEO SAR
		5	**Magnetospheric Constellation**
		6	Ionospheric Mappers
2	Inner solar system, simple profile, moderate ΔV	7	Space Interferometry Mission (SIM)
		8	Starlight ST-3
3a	Inner solar system, sample return	9	Comet Nucleus Sample Return (CNSR)
		10	**Mars Sample Return (low thr)**
			Mars Sample Return (hi thr)
3b	Inner solar system, complex profile, mod. to high ΔV (Lagrange point missions will be considered as complex due to the sensitivity of the trajectory to perturbations. ** = E-S, E-M L point missions)	11	EASI**
		12	Pole Sitter**
		13	Sub L1 point mission**
		14	**Solar Sentinels****
		15	**Solar Polar Imager**
		16	NGST**
		17	Terrestrial Planet Finder**
		18	Outer Zodiacal Transfer**
4	Outer solar system, simple profile, high ΔV		Outer Zodiacal Transfer
5	Outer solar system, complex profile, includes propulsion in the outer solar system	19	**Titan Organics Explorer Orbiter & Lander****
		20	**Neptune Orbiter (AP)**
			Neptune Orbiter (AC)
		21	**Europa Lander**
		22	Solar Probe
6	Beyond outer solar system	23	**Interstellar Probe****
7	HEDS lunar, cislunar, & Earth vicinity	24	**Moon & Earth-Moon Libration Points**
		25	Sun-Earth Libration Points
8	HEDS Asteroids / Mars vicinity	26	Near Earth Asteroids
		27	**Mars Cargo & Piloted**

In-space ISTP Science & Exploration Missions
Propulsion Options Currently Under Study

#	Option	Center
1, 2	SOA Chemical / Adv. Chemical	(MSFC)
3, 4	Electric Propulsion (Ion), Nuclear, Comb. Solar/Chem	(JPL/GRC/MSFC)
5, 6	Electric Propulsion (Hall) N, S/Ch	(JPL/GRC/MSFC)
7, 8	Electric Propulsion (PIT) N, S/Ch	(JPL/GRC/MSFC)
9, 10	Electric Propulsion (MPD) N, S/Ch	(JPL/GRC/MSFC)
22	Radio-Isotope Electric Propulsion	(GRC)
11	VaSIMR	(MSFC/JSC)
12	Solar Thermal Propulsion	(MSFC)
13	Nuclear Thermal Propulsion	(GRC)
14	Bimodal Nuclear Thermal Propulsion	(GRC)
21	Hybrid NTP/NEP	(GRC)
15	M2P2 (Mag-sail)	(MSFC)
16	Solar Sails	(JPL)
17	MX Tether/Augmentation	(MSFC/TU)
18	ED Tether/Augmentation	(MSFC/SRS)
19	Fusion	(GRC/MSFC)
20	Aeroassist	(LaRC)

— All with or without: a) CFM, b) ISPP, and c) LW components as makes sense.

In-space ISTP Science & Exploration Missions
General IISTP Mission Groundrules & Assumptions

- Mass Margins:
  - Systems Team Spacecraft: 30%
  - Customer-supplied S/C: 0% (use as given by customer)
  - Propulsion: 10-30% based on TRL
  - SEP (JPL):   30% for both 5 kW & 10 kW Ion
  - Launch Vehicle Performance: 10%

- On-board Propulsion:
  - Stage sizing percentage: 28%*

  *(For cases that need other non-high thrust stages - for example MSP2 will need a capture stage for the Neptune Orbiter mission. This includes a lot of stuff - tanks, engines, power, thermal, etc - just for the stage. This is intentionally high - to consider the small scale the robotic missions use.)

- ΔV Margins:
  - Dispersions/Performance/Isp margin: 2%
  - Gravity Losses: should include

- Tank Fractions:
  - Primary LOx/LH$_2$ prop systems: 6½% (for MR=5.5, 4% LOx; 19% LH$_2$)
  - Primary storable prop systems: 5%
  - Supercritical propellant systems: 5% for Xenon, 7.5% for Krypton
  - Liquid Inert systems: 2.5% for Xenon, 0.5% for Krypton

Neptune Orbiter Package

IISTP Phase I Final Report
September 14, 2001

In-space ISTP Science & Exploration Missions
Neptune Orbiter Mission Groundrules & Assumptions

- Tour/Mission Duration at Neptune: ~3 years on station
- Neptune Spacecraft/bus Payload Complement:
  - All-propulsive (AP) case: 500 kg S/C
  - Aerocapture (AC) case: 500 kg S/C + 75 kg mono-prop = 575 kg total
- Aeroassist:
  - Neptune: 30% of braked mass (which includes the aeroshell/ballute)
    - (Note: a 575 kg payload requires a 246 kg aeroshell/ballute)
- Trajectories
  - High Thrust AC: 10 years out to destination
  - SEP low thrust from escape: 10, 11 years out to destination
  - NEP Low Thrust: 1 year spiral + 10 years out to destination
  - M2P2 AC: 10 years
  - M2P2 AP: 12 years
  - High Thrust NTP AP: 12 years
  - High Thrust Chem AP: 14 years
- Parking Orbits (PO)
  - Initial AC PO @ Triton: 100 km (125 km ref?) by ~330,000 km (both altitudes)
    (requires additional mono-propellant system to do the 175 m/s periapsis raise maneuver)
  - Initial AP PO @ Triton: 6191 km by ~330,000 km (both alt.)
  - Final PO @ Triton: 6191 km by ~330,000 km (both alt.)

Larry Kos/TD80
4/4/2001

In-space ISTP Science & Exploration Missions
Propulsion System Groundrules & Assumptions

- **SOA Chemical/Adv. Chemical Propulsion:**
  - Resultant stage mass fractions: ~0.85 (primary), 0.78 (on-board prop)
  - Propellant Specific Impulses (Isp): 465 (LOx/LH$_2$), 325 (storable) sec
  - Adv. Chemical Propellant Isp: 377 sec
- **Electric Propulsion (Hall, Ion, MPD, PIT, etc.):**
  - EP Stage Overall Specific Mass, α: 30-33 (NEP), 35-40 (SEP)
  - EP Stage Power, (kWe or MW): 20 - 100 kWe (NEP), 24 (SEP)
  - EP Power Specific Mass, (kg/kW): 77 - 26 (NEP), 7 (SEP)
  - EP Thruster Specific Mass, (kg/kW): 2 - 9 (NEP), 6 (SEP)
  - Propellant Specific Impulse: 9000 sec (NEP), 4000-5000 (SEP)
- **VaSIMR**
  - EP Stage Overall Specific Mass, α: 30 (NEP), x (SEP)
  - Isp: 4000 sec (low), 15,000 sec (high); Power: 100 kWe (NEP), x kWe (SEP)
- **Solar, Nuclear & Bimodal Thermal Propulsion, Fusion:**
  - Isp: ; Mass Fraction or dry mass:
- **Sails:**
  - Solar Sail
    - Areal Density: 10 g/m² (2010 Launch) and 5 g/m² (2015 Launch) (includes sail film, structure, residual deployment mechanism (that can't be jettisoned), and a boom & gimbals between the sail and the payload for center-of-mass vs. center-of-pressure yaw & pitch attitude control)
    - Area: 90,000 m² for a 300 m square sail side ('10 launch) & 200,000 m² for a 447 m square sail ('15 launch)
    - Sail Deployment Mechanism (jettisoned after deployment) = 25% of Sail mass
  - Mag Sail (M2P2) Power: 7-12 kW; Thrust: 1 N/kW
- **Tethers**
  - Base station mass: 120 mt; Tether length: 140 km; Max ΔV: 3.07 km/s

IISTP Systems Team
4/4/2001

IISTP Mission Analysis Considerations for CFM

Neptune Orbiter Mission

- Propellant Characteristics:
  - Type of propellant — LOx/LH$_2$, NTO/MMH, & LH$_2$, Xe
  - Usage profile — Use LOx/LH$_2$ in hrs
  - Quantities (fuel and oxidizer) — 1.3 + 6.9 = **8.2** mt (@ SOA chem MR = 5.5)
  - Number of tanks, if known — 1 each
  - Fairing dimensions for tank (OD, length) — fits in Delta IV shroud
- Mission Characteristics:
  - Duration — ~13-15 years
  - Environments — Neptune vicinity
  - Solar constant — 2 - 1/900 of Earth
  - Time between burns — hours (2-burn TNI & Neptune vicinity)
  - Number of burns — 1-2
- Other Considerations / Requirements:
  - Loss tolerance vs. zero losses — zero losses if poss.
  - Restrictions
  - Ground ops — LOx/LH$_2$, MMH/NTO, LH$_2$, Xe Ops
  - On orbit refueling — none
  - Propellant use at destination:
    - Orbit capture — storable bi-prop
    - Station keeping — storable mono-prop

Steve Tucker / MSFC
Dave Plachta / GRC
Larry Kos/TD30
4/4/2001

Baseline/Pivot & Proposed Missions
Systems/Mission/Sizing Analyses Results

Larry Kos/TD30
3/12/2001

IISTP Science & Exploration Missions
Chem/Aero Propulsion Options to Neptune/Triton

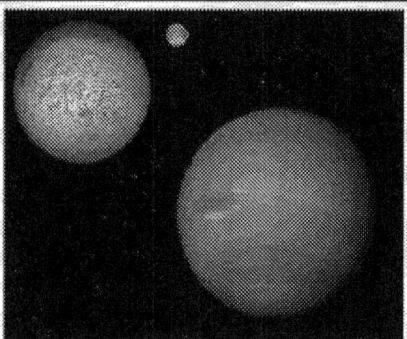

- **Assessment Results**
  - IMGTO: 11 mt (AP), 9.5 mt (CH_4), 9.5 mt (AC)
    (114 mt for 12 yr AP 400 km circ, 37 mt for 14 yr AP 400 km circ)
  - Departure (AP): April 8, 2010
  - Neptune Arrival (AP): May 4, 2024
  - Departure (AC): April 13, 2010
  - Neptune Arrival (AC): May 20, 2020
  - Mission Duration (AP): 14 yrs + tour, ~17 yrs
  - Mission Duration (AC): 10 yrs + tour, ~13 yrs
  - Total Mission Operation Time: ~13-15 yrs
  - No. of Launches (AP & CH_4): 1 (Delta IV H)
  - No. of Launches (AC): 1 (Delta IV H)

- **Transportation Approach**
  - Off-the-shelf LOx/LH, RL-10 B-2 engine, I_{sp} = 465 sec, Thrust = 24,750 lb., 11% scaling mass, 0.5 mt constant, perform 1-burn maneuver from GTO, plus 325 (377) sec Isp, 904 (690) kg wet mass storable system for propulsive braking at Neptune in the all-propulsive (AP) case
  - Ballute or aeroshell, for aerocapture case (AC), scaling relation/fraction is 30% for V^∞ = 11.7 km/s giving 246 kg, and includes the 75 kg periapsis raise mono-propellant system.

- **Issues**
  - Larger launch vehicle required
  - Ballute or aeroshell can be used high entry speeds (24 - 29 km/s, V^∞ ~5.5 - 11.7)
  - Atmospheric characterization for Neptune
  - 2nd onboard propulsion system required for either Neptune propulsive capture in the all-prop case, and for the periapsis raise after capture in the aerocapture case.

IISTP Science & Exploration Missions
Tether/Chem/Aero Propulsion Options to Neptune/Triton

- **Assessment Results**
  - IMLEO: 9.7 mt (AP), 8.4 mt (AC)
  - Departure (AP): April 8, 2010
  - Neptune Arrival (AP): May 4, 2024
  - Departure (AC): April 13, 2010
  - Neptune Arrival (AC): May 20, 2020
  - Mission Duration (AP): 14 yrs + tour, ~17 yrs
  - Mission Duration (AC): 10 yrs + tour, ~13 yrs
  - Total Mission Operation Time: ~15 yrs
  - No. of Launches (AP): 1 (Delta IV M+4,2; or a SL)
  - No. of Launches (AC): 1 (Delta IV M+5,2; or a Zenit 3SL-Sea Launch)

- **Transportation Approach**
  - Same chemical propulsion as the chemical case
  - Same aeroassist as the chemical case
  - 10 mt PYLO tether base station size ... see the "four sizes fits all" tether basestations (pages 11 through 14).

- **Issues**
  - ~120 mt base station in LEO, 0° required
  - Ballute or aeroshell as for the chemical case
  - Atmospheric characterization for Neptune
  - 2nd onboard propulsion system required for either Neptune propulsive capture in the all-prop case, or for the periapsis raise after capture in the AC case.

MXER Tether Boost Station for IISTP
(10 MT payload capability)

Assessment Results
- IMLEO: 120,000 kg + 21,000 kg spent upper stages
- Operational Orbit: 400 x 13,000 km
- Tether Length: 140 km
- Launch requirement: 6 Delta 4H

Transportation Approach
- Momentum exchange/electrodynamic reboost (MXER) tether facility in Earth orbit boosts spacecraft to high-energy, pre-escape trajectories
- High-thrust propulsion (chemical or NTR) conducts DV at perigee to target hyperbolic C_3
- Low-thrust propulsion (SEP, NEP, sails) uses lunar swingby to achieve low C_3 heliocentric orbit
- MXER tether facility supports commercial GEO missions as well as interplanetary spacecraft

Issues
- The MXER tether system must be developed and in place (no capability of this type currently exist)
- Development/demonstration of key MXER tether technology and systems is required
- Tether boost timing adds an additional factor to consider in designing and optimizing robust interplanetary trajectories
- Tether facility operations
- Large mass associated with tether (8-10x payload mass)
- Equatorial launch requirement

Rob Hoyt / Tethers Unlimited
Kirk Sorensen/TD40
3/23/2001

IISTP Science & Exploration Missions
Nuclear Thermal Propulsion Options to Neptune/Triton

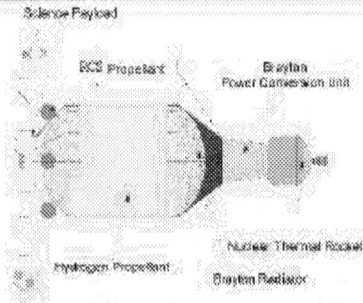

Assessment Results
- IMLEO: 6271 kg in 407 km circular orbit
- 748 kg delivered to 125 km by 330000 km Neptune orbit
- Departure (AC): April 10, 2010
- Neptune Arrival (AC): June 13, 2020
- Mission Duration (AC): 10 yrs +tour, ~13.2 yrs
- No. of Launches: 1 Delta IV M

Issues
- Spacecraft is volume constrained on a Delta III with 2600 kg of additional mass capability from the ELV
- LOX augmentation may reduce tank size sufficiently to fit on a Delta III
- Ballute or aeroshell can be used high entry speeds (24 - 29 km/s, V_{inf} 11.6 km/sec).
- Atmospheric characterization for Neptune
- 2nd onboard propulsion system required for the periapse raise after capture in the aerocapture case.

Transportation Approach
- MITEE NTRE : 27.4 kN Thrust @ 940 s Isp
- Carbon Composite tank sized for Delta IV 5m Shroud
- NTR used for Trans-Neptune Injection & jettisoned thereafter
- Science payload 500 kg plus 75 kg monoprop system for periapsis raise at Neptune
- Ballute or aeroshell for aerocapture case (AC), scaling relation/fraction: 30% for V* = 11.6 km/s giving 173 kg

Leo Dudzinski / GRC
3/29/2001

IISTP Phase I Final Report
September 14, 2001

IISTP Science & Exploration Missions
Bimodal Nuclear Thermal Propulsion Options to Neptune/Triton

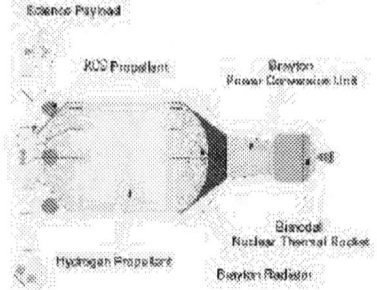

- **Assessment Results**
  - IMLEO: 12,304 kg in 407 km circular orbit
  - 3000 kg delivered to 6191 km by 330,000 km Neptune orbit
  - Departure (AP): April 8, 2010
  - Neptune Arrival (AP): April 27, 2022
  - Mission Duration (AP): 12 yrs + tour, ~15 yrs
  - No. of Launches: 1 Delta IV M+ (5,4)

- **Issues**
  - No aeroshell required at Neptune
  - No atmospheric characterization for Neptune required
  - No 2nd onboard propulsion system required for the periapsis raise after capture in the aerocapture case.

- **Transportation Approach**
  - MITEE NTRE : 27.4 kN Thrust at 940 sec Isp
  - Bimodal reactor produces 2.9 kWe for S/C power (400We) and cryogenic propellant refrigeration to maintain zero boil-off
  - Carbon Composite tank sized for Delta IV 5m Shroud
  - NTR used for Trans-Neptune Injection & Neptune orbit capture. Bimodal power used throughout mission.
  - Science payload 500 kg

IISTP Science & Exploration Missions
SEP/ NTP/Aero Hybrid Option to Neptune/Triton

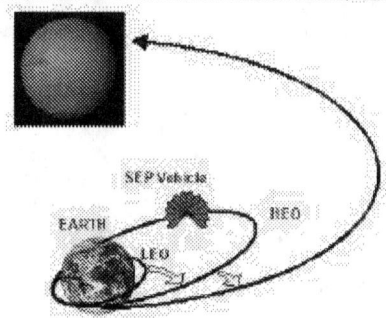

- **Assessment Results**
  - IMLEO: 6.6 mt @ 400 km
    - 2.4 mt SEP, 3.4 mt NTR
    - 0.3 mt AB, 0.5 mt S/C
  - Launch: Mar 7, 2010
  - Earth Departure: Apr 11, 2011
  - Neptune Arrival: Apr 10, 2024
  - Mission Duration: 14 yrs + tour, ~17 yrs
  - Total Mission Operation Time: ~17 yrs
  - No. of Launches: 1 (Atlas V 500 or Delta III)

- **Issues**
  - The required 10 kW Hall thruster needs to be flight qualified and life tested
  - Low cost array technology needed to make over all mission cost attractive
  - Advanced attitude control system technology development required to steer spacecraft during elliptical spiral transfer
  - Radioisotope power system required on spacecraft for deep space power and NTP system used for escape maneuver. Potential safety issues during elliptical spiral.

- **Transportation Approach**
  - Transfer from LEO (400 x 400 km) to a HEO (2500 x 120,500 km) using a 30 kW BOL (25 kW EOL) solar powered Hall thruster system
  - Solar electric system separates for spacecraft/chemical system
  - Escape Earth to C_3 of 155 km^2/s^2 using nuclear thermal rocket system
  - Spacecraft captures in to Neptune system using aerocapture system

IISTP Science & Exploration Missions
SEP/Chem (10 kW Hall) Hybrid Option to Neptune/Triton

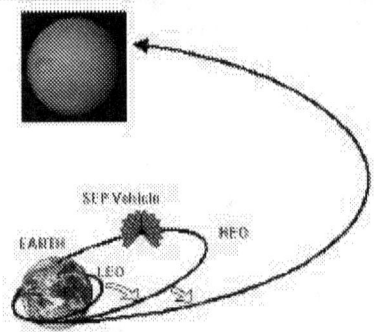

- **Assessment Results**
  - IMLEO: 12.7 mt @ 400 km
    - 4.6 mt SEP, 7.3 mt Solids
    - 0.3 mt AB, 0.5 mt S/C
  - Launch: Mar 7, 2010
  - Earth Departure: Apr 11, 2011
  - Neptune Arrival: Apr 10, 2024
  - Mission Duration: 14 yrs + tour, ~17 yrs
  - Total Mission Operation Time: ~17 yrs
  - No. of Launches: 1 (Atlas V 520 or Delta IV M+ (5,4))

- **Transportation Approach**
  - Transfer from LEO (400 x 400 km) to a HEO (800 x 120,500 km) using a 50 kW BOL (42 kW EOL) solar powered Hall thruster system
  - Solar electric system separates for spacecraft/chemical system
  - Escape Earth to C_3 of 155 km²/s² using dual solid rocket motor system
  - Spacecraft captures in to Neptune system using aerocapture system

- **Issues**
  - The required 10 kW Hall thruster needs to be flight qualified and life tested
  - Low cost array technology needed to make over all mission cost attractive
  - Advanced attitude control system technology development required to steer spacecraft during elliptical spiral transfer
  - Radioisotope power system required on spacecraft for deep space power. Potential safety issues during elliptical spiral.

IISTP Science & Exploration Missions
NEP/Chem/Aero (10 kW Hall) Option to Neptune/Triton

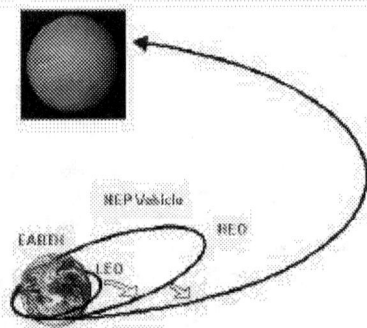

- **Assessment Results**
  - IMLEO: 13.8 mt
    - 5.7 mt NEP, 7.3 mt Solids
    - 0.3 mt AB, 0.5 mt S/C
  - Launch: Mar 7, 2010
  - Earth Departure: Apr 11, 2011
  - Neptune Arrival: Apr 10, 2024
  - Mission Duration: 14 yrs + tour, ~17 yrs
  - Total Mission Operation Time: ~17 yrs
  - No. of Launches: 1 (Atlas 530)

- **Transportation Approach**
  - Transfer from LEO (800 x 800 km) to a HEO (800 x 120,500 km) using a 50 kWe space nuclear reactor powering Hall thruster system
  - Solar electric system separates for spacecraft/chemical system
  - Escape Earth to C_3 of 155 km²/s² using dual solid rocket motor system
  - Spacecraft captures in to Neptune system using aerocapture system

- **Issues**
  - Public/leadership perception of nuclear environmental issues
  - Development of necessary nuclear power system ground testing infrastructure
  - The required 10 kW Hall thruster needs to be flight qualified and life tested
  - Nuclear power system has associated safety issues for launch and during spiral.
  - Advanced attitude control system technology development required to steer spacecraft during spiral transfer

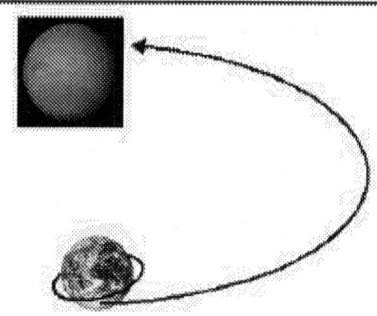

IISTP Science & Exploration Missions
SEP/Aero/Chem (10-kW Ion 5000 sec) Option to Neptune/Triton

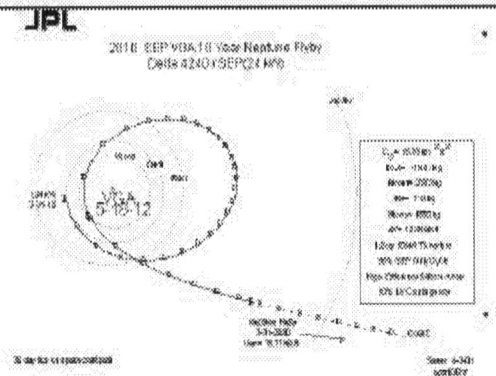

2010 SEP VGA1β Vinf Neptune Flyby
Delta 4240 rSEP(24 kW)

- **Assessment Results**
  - IMEsc: 2.58 mt
  - Xe prop. Mass: 0.71 mt
  - Power: 24 kW
  - Departure: March 31, 2010
  - Neptune Arrival: March 31, 2020
  - Trip time: 10 yrs
  - Total Mission Ops Time: ~13 yrs
  - No. of Launches: 1 (Delta 4240)

- **Issues**
  - Grids development
  - High voltage propellant isolator development
  - Light weight engine body
  - 550-kg throughput demonstration
  - Wide 1-10 kW range throttling
  - Multi-engine demonstration
  - New PPU & DCIU development
  - Some advanced feed system components
  - Aeroshell/ballute mass fraction for high Vinf

- **Transportation Approach**
  - Launch to C3 = 15.7 km²/s²
  - Venus GA 2 yrs after launch
  - SEP module jettisoned at ~ 3.5 AU
  - SEP module has 2+1 spare eng, 2+1 spare PPUs
  - Ballute or Aeroshell scaling fraction= 30%, V' @ Neptune = 15.8 km/s
  - Payload mass: 500 kg S/C, 246 kg Aeroshell, 75 kg Mono-prop.

M. Hood/JPL
4/9/2001

IISTP Science & Exploration Missions
Solar Sail/Aero/Chem Option to Neptune/Triton

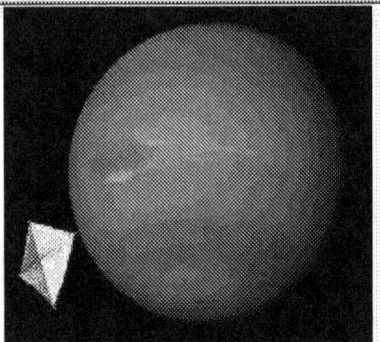

- **Assessment Results**

IMEsc:	5.22 mt	5.22 mt
Areal Dens (g/m²)	10	5
Square Sail Side (m)	593	839
• Tech Projection (m)	300	447
Departure:	2010	2015
Trip Time:	9.2 yrs	5.3 yrs
Neptune Arrival:	2019	2020
Mission Duration:	12.2 yrs	6 yrs
Total Mission Ops Time:	12.2 yrs	6 yrs
No. Launches:(Atlas V 550)	one	one

(Atlas V 550 = 5.22 mt w/ 10% L/V margin)

- **Transportation Approach**
  - Launch Vehicle delivers to C_3 = 0
  - Sail deployed
    - Sail deployment mech = 25% of Sail mass (jettisoned after deployment)
  - Sail flies near sun to build up speed (higher light pressure) - Rmin = 0.3 AU
  - Sail jettisoned ~ 5 AU
  - At this point, mission similar to Chem/Aero option: S/C Aerocaptures into Neptune orbit
    - 246 kg aeroshell/ballute + 75 kg periapsis raise monoprop system + 500 kg net spacecraft

- **Issues**
  - Sail size required exceeds technology projections
  - Volumetric storage of sail in L/V payload fairing
  - Achievable Sail areal density, size
  - Control, stability, dynamics
  - Aeroshell/ballute system mass for short trip times (high V_{inf})

Robert Frisbee (JPL)
3/20/2001

IISTP Science & Exploration Missions
NEP (Ion) Option to Neptune/Triton

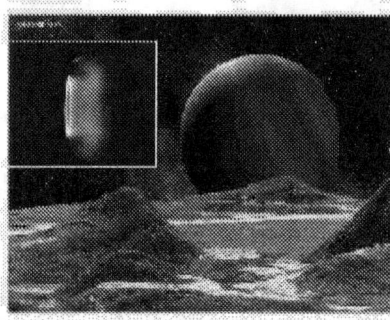

- **Assessment Results**
  - IMLEO: 7.9 mt
  - Departure: September 20, 2010
  - Neptune Arrival: September 20, 2020
  - Mission Duration: ~1yr +10 yr + tour, ~14 yrs
  - Total Mission Operation Time: ~14 yrs
  - No. of Launches: 1 (Delta IV M+)

- **Issues**
  - Public/leadership perception of nuclear environmental issues
  - Development of necessary nuclear power system ground testing infrastructure
  - The required 10-30 kW thrusters need to be be flight qualified and life tested
  - Nuclear power system has associated safety issues for launch and during spiral.
  - Advanced attitude control system technology development required to steer spacecraft during spiral transfer

- **Transportation Approach**
  - Depart from 1000 km circular about Earth, 10 year TOF in heliocentric space, spiral down to 100,000 km circular altitude about Neptune (~ capture ΔV)
  - Vehicle Parameters: Payload = 500 kg, Propulsion System α = 30 kg/kW, Tankage Fraction = 10% of Propellant, Overall Efficiency = 60%, Power = 100 kW, Optimal Isp ~ 9000 seconds

IISTP Science & Exploration Missions
NEP (MPD) Option to Neptune/Triton

- **Assessment Results**
  - IMLEO: 8.6 mt
  - Departure: September 20, 2010
  - Neptune Arrival: September 20, 2020
  - Mission Duration: ~1yr +10 yr + tour, ~14 yrs
  - Total Mission Operation Time: ~14 yrs
  - No. of Launches: 1 (Delta IV M+)

- **Issues**
  - Public/leadership perception of nuclear environmental issues
  - Development of necessary nuclear power system ground testing infrastructure
  - The required 100 kW thruster needs to be be flight qualified and life tested
  - Nuclear power system has associated safety issues for launch and during spiral.
  - Advanced attitude control system technology development required to steer spacecraft during spiral transfer

- **Transportation Approach**
  - Depart from 800 km circular about Earth, 10 year TOF in heliocentric space, spiral down to 100,000 km circular altitude about Neptune (~ capture ΔV)
  - Vehicle Parameters: Payload = 500 kg, Propulsion System α = 33 kg/kW, Tankage Fraction = 10% of Propellant, Overall Efficiency = 65%, Power = 100 kW, Optimal Isp ~ 9000 seconds

IISTP Science & Exploration Missions
NEP (VaSIMR) Option to Neptune/Triton

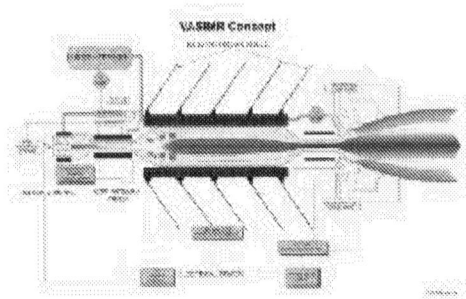

- **Assessment Results**
  - IMLEO: 8.4 mt
  - Departure: September 18, 2010
  - Neptune Arrival: September 15, 2020
  - Mission Duration: ~1/2 yr +10 yr +tour, ~13.5 yrs
  - Total Mission Operation Time: ~13.5 yrs
  - No. of Launches: 1 (Delta IV M+)

- **Issues**
  - Public/leadership perception of nuclear environmental issues
  - Development of necessary nuclear power system ground testing infrastructure
  - The required 10-50 kW thrusters need to be flight qualified and life tested
  - Nuclear power system has associated safety issues for launch and during spiral.
  - Advanced attitude control system technology development required to steer spacecraft during spiral transfer

- **Transportation Approach**
  - Depart from 800 km circular about Earth, 10 year TOF in heliocentric space, spiral down to 100,000 km circular altitude about Neptune (~ capture ΔV)
  - Vehicle Parameters: Payload = 500 kg, Propulsion System α = 30 kg/kW, Tankage Fraction = 10% of Propellant, Overall Efficiency = 60%, Power = 100 kW, Isp ~4000 & 15,000 seconds

Larry Kos/TD30
3/28/2001

IISTP Science & Exploration Missions
Mag-sail/Aero/Chem Option to Neptune/Triton

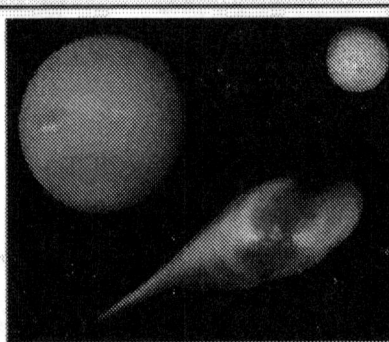

- **Assessment Results**
  - IMEsc: 2.6 mt (AP), 1.5 mt (AC)
  - Departure (AP): X, 2010
  - Neptune Arrival (AP): X, 2024
  - Departure (AC): X, 2010
  - Neptune Arrival (AC): X, 2020
  - Mission Duration (AP): 14 years + tour, ~17 yrs
  - Mission Duration (AC): 10 years + tour, ~13 yrs
  - Total Mission Operation Time: ~13-17 yrs
  - No. of Launches (AP): 1 (~Delta III)
  - No. of Launches (AC): 1 (~Delta II)

- **Transportation Approach**
  - Launch Vehicle delivers to C_3 = 0
  - Thrust = 1 N/kW
  - Propellant Consumption = 1 kg/kW/day
  - AC payload = 575kg + 246kg (AC)
  - AP payload = 500kg + 904kg (Chem Stg)
  - Solar Power Sys.
    - α = 30
    - AC Power = 7.4 kW
    - AP Power = 12.7 kW

- **Issues**
  - Demonstration of the physical principles associated with this propulsion concept & flight experience
  - Effectiveness of this concept to provide useful space transportation capabilities (possible space environment interaction availability limitations)
  - Demonstration of such supporting technologies and a system that can meet system operation and duty cycle life conditions
  - Vehicle system design integration & operation issues

Tara Poston/TD30
3/16/2001

IISTP Science & Exploration Missions
Fusion Option to Neptune/Triton

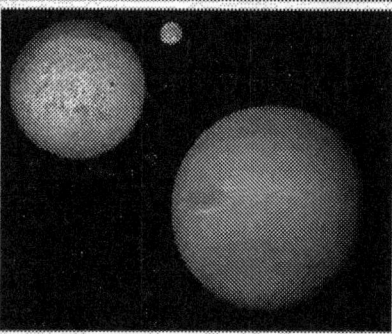

- **Assessment Results**
  - IMLEO: TBD mt
  - Departure: TBD
  - Neptune Arrival: TBD
  - Mission Duration: TBD d +tour, TBD yrs
  - Total Mission Operation Time: TBD yrs
  - No. of Launches: TBD (... LV ...)

- **Issues**
  - Time frame a fusion system can be ready is uncertain
  - Uncertainty in most parameters for estimating system performance
  - Possible "disruption" of Earth's magnetosphere with fusion exhaust
  - Options still include both inertial electrostatic confinement (IEC) and magnetized target fusion (MTF) as well as others

- Transportation Approach
  - Fusion system, Isp: 10,000 - 100,000 sec, Propellant mass fraction: TBD, Tank Fraction: TBD
  - Departure orbit: TBD (possibly 2500 km alt, or possibly required to start well above GEO)

Trajectory / Mission Plots

IISTP Phase I Final Report
September 14, 2001

Neptune/Triton Chemical Propulsion Option
Mission Event Sequence

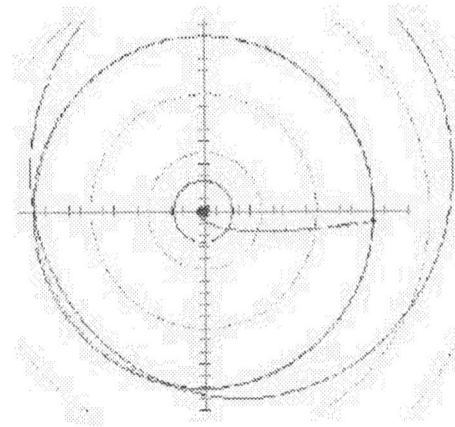

Propulsive Capture Trajectory

- (Chemical) Mission Approach:
  - Delta IV-H LV places 500 kg payload with chemical stage into GTO orbit
  - LOx/LH$_2$ chemical propulsion system performs 6.3 km/s ?V to inject onto 14 year coast
  - Drop LOx/LH$_2$ stage after burn
  - Propulsively capture with 2.2 km/s ?V using storable propellant to capture into 6200 km x 330,000 km altitudes (1.25 R$_N$ x 14.2 R$_N$)
  - Begin three year tour of Neptunian system, utilizing swingbys of Triton to modify tour as needed/as possible

Neptune/Triton Chemical Propulsion Option
Mission Event Sequence

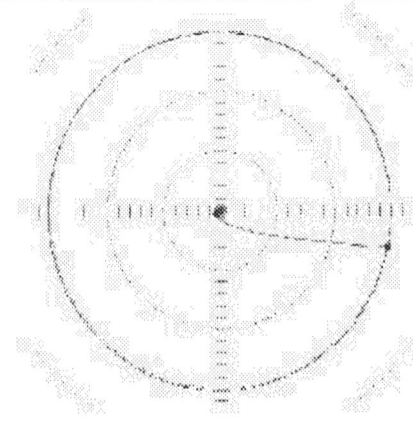

Aerocapture Trajectory

- Chemical/Aeroassist Mission Approach:
  - Delta IV-H LV places 575 kg payload with chemical stage into GTO orbit
  - LOx/LH$_2$ chemical propulsion system performs 7.0 km/s ?V to inject onto 10 year coast
  - Drop LOx/LH$_2$ stage after burn
  - Capture with aeroassist (26 km/s entry speed) using aeroshell or ballute to capture into 100 km x 330,000 km altitudes (1.00 R$_N$ x 14.2 R$_N$)
  - Coast to first apoapsis and use on-board mono-propellant system to raise periapsis to 1.25 R$_N$
  - Begin three year tour of Neptunian system, utilizing swingbys of Triton to modify tour as needed/as possible

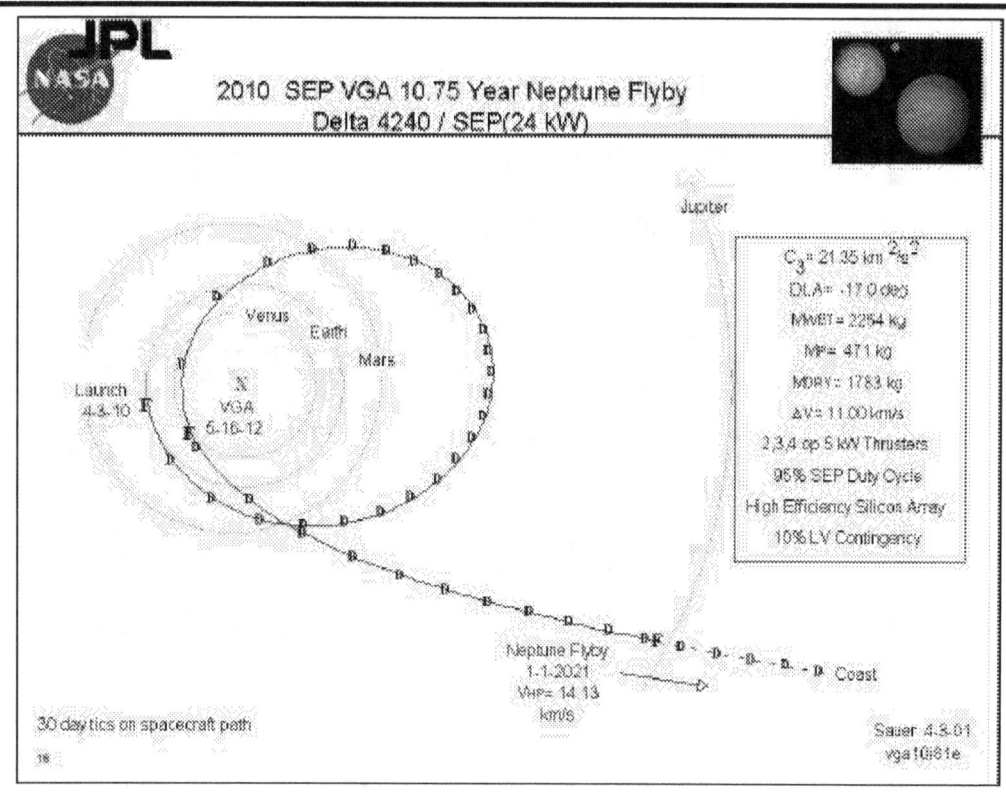
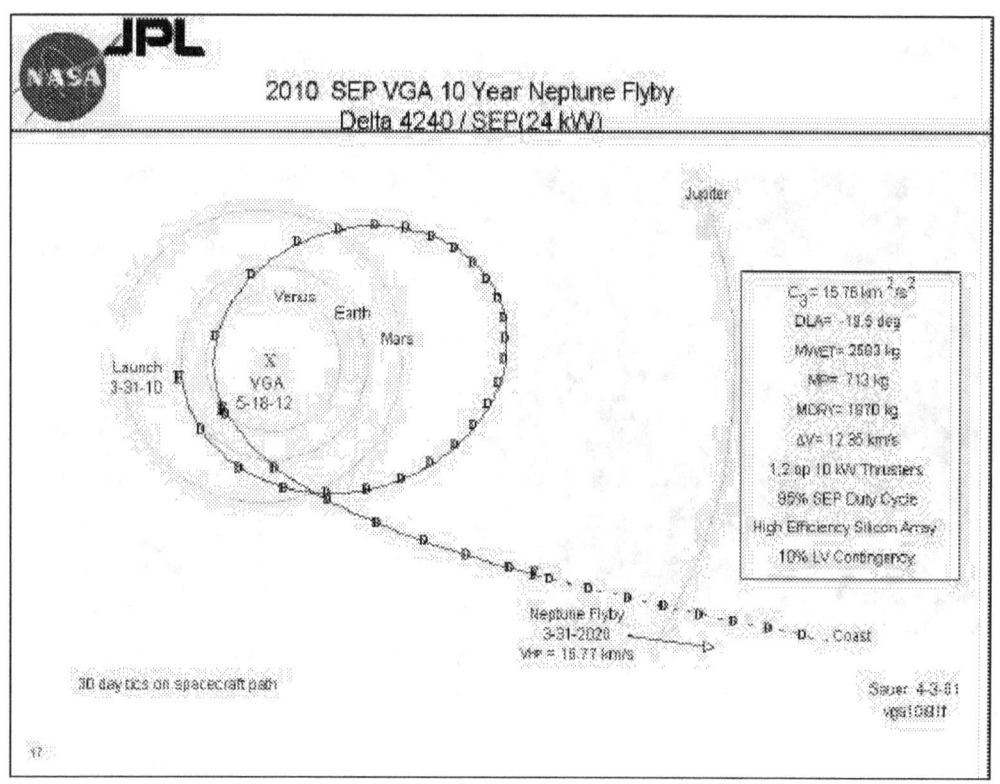

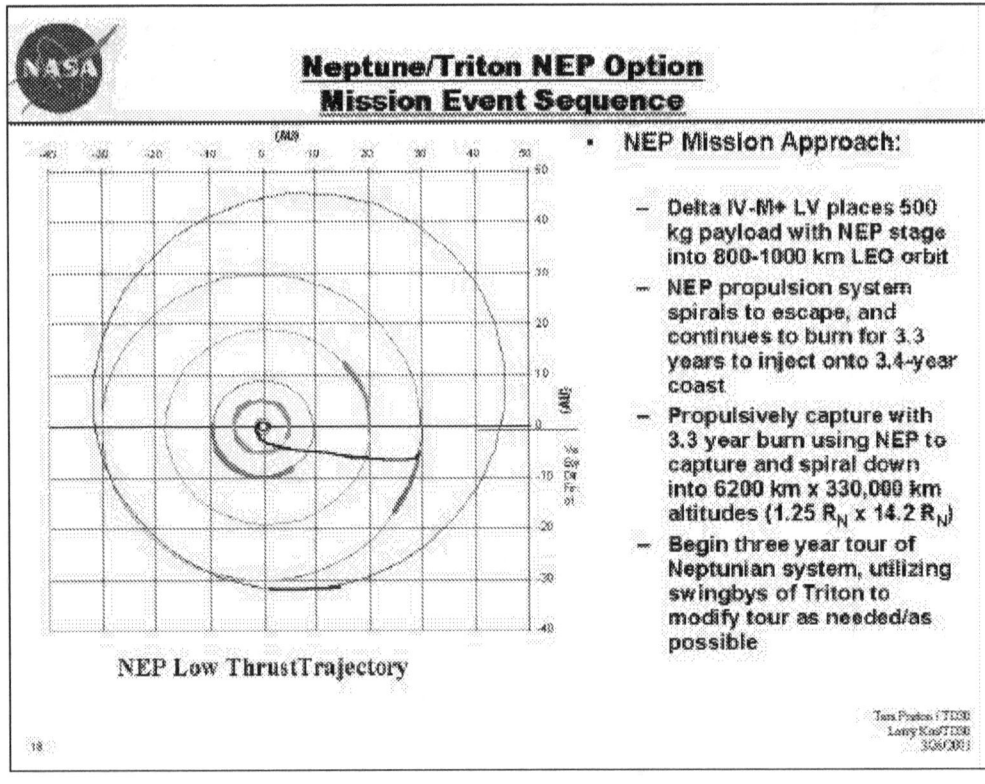

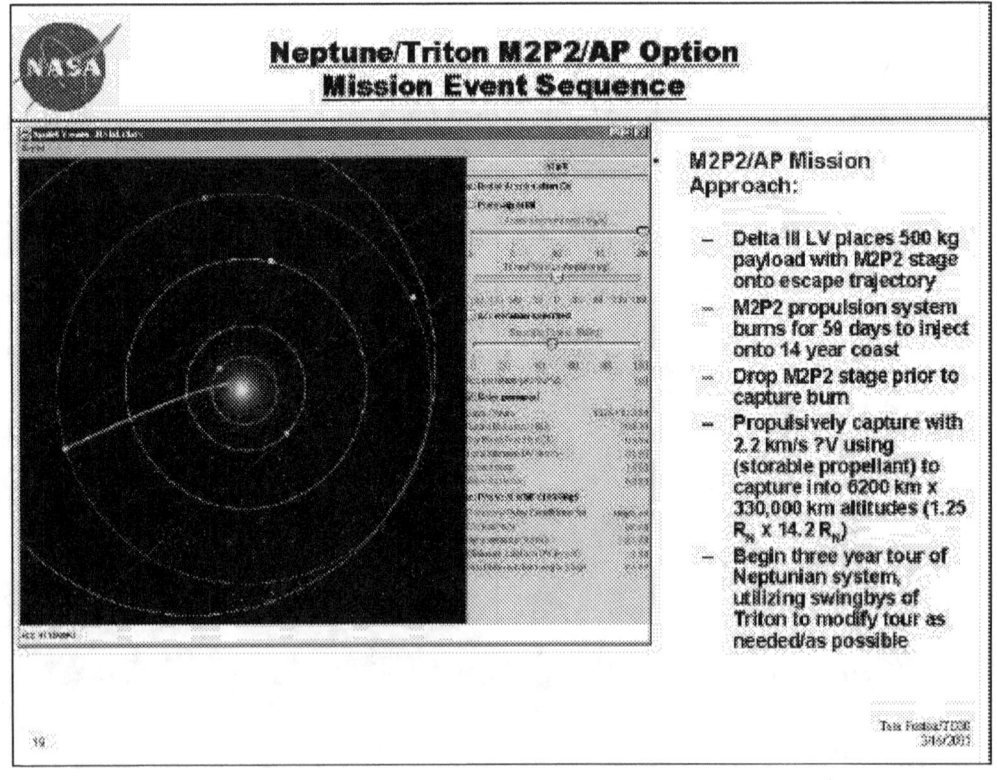

Neptune/Triton M2P2/AC Option
Mission Event Sequence

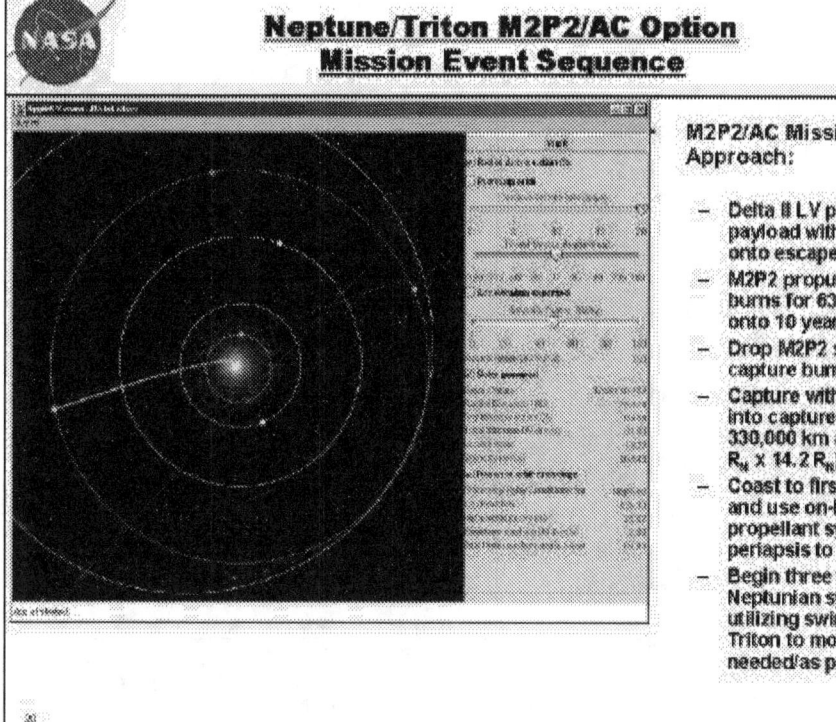

M2P2/AC Mission Approach:

- Delta II LV places 575 kg payload with M2P2 stage onto escape trajectory
- M2P2 propulsion system burns for 63 days to inject onto 10 year coast
- Drop M2P2 stage prior to capture burn
- Capture with aeroassist into capture into 100 km x 330,000 km altitudes (1.00 R_N x 14.2 R_N)
- Coast to first apoapsis and use on-board mono-propellant system to raise periapsis to 1.25 R_N
- Begin three year tour of Neptunian system, utilizing swingbys of Triton to modify tour as needed/as possible

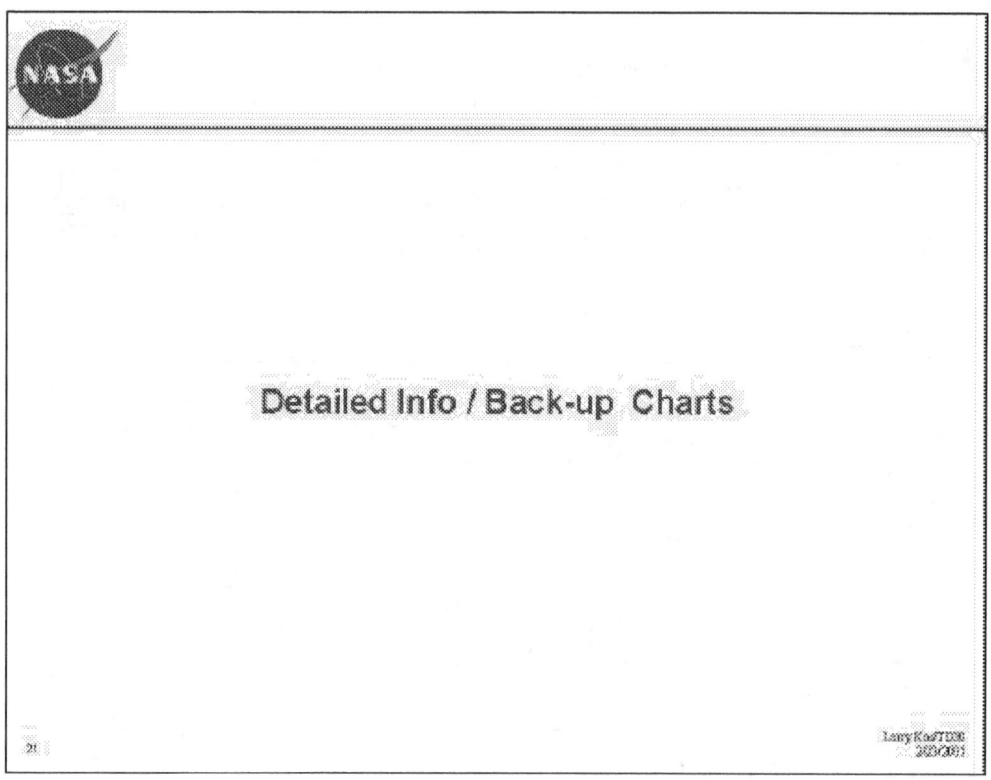

Detailed Info / Back-up Charts

In-space ISTP Activity Triage Worksheet
Proposed Missions (28) and Mission Categories (8)

Launch Vehicle Performance Table
(AIAA/Isakowitz, 3rd Edition) For IISTP LV selection

Destination: Launch Vehicle	COST, 2001 $M	400 km circ, 28.5°	800 km circ, 28.5°	400 km circ, 51.6°	GTO (167 x 35, 185 km, i=27°/7°)	C₃ = 0 km²/s²	C₃ = x.x km²/s²
*Athena II	~$25	1,755	1,315	1,685	525	348	N/C
*Atlas V 500	~$90	~9,080	~8,700	~8,600	4,051	~2,800	See
*Atlas V 510	~$95	~10,600	~10,300	~10,000	4,849	~3,400	LeonLVs
*Atlas V 400	~$88	~11,080	~10,600	~10,500	4,948	~3,500	worksheet
*Atlas V 520	~$100	~12,400	~11,900	~11,600	5,945	~4,200	in
*Atlas V 530	~$106	~15,300	~14,800	~14,500	6,841	~4,900	this
*Atlas V 540	~$111	~16,700	~16,100	~15,800	7,539	~5,400	work
*Atlas V 550	~$116	~17,900	~17,300	~16,900	8,137	~5,800	book
Delta II 7325/20	~$53	2,500	2,250	2,435	1,000	675	(See
Delta II 7425/20	~$53	2,800	2,550	2,715	1,130	780	Isakowitz
Delta II 7925/20	~$58	4,660	4,100	4,440	1,870	1,230	page 106)
Delta III	~$88	8,000	7,550	7,300	3,810	2,700	
Delta IV Med	~$88	~7,880	~7,000	7,700	3,900	~2,600	TBD
Delta IV M+(5,2)	~$98	~9,380	~8,400	9,180	4,350	~2,900	TBD
Delta IV M+(4,2)	~$98	~10,600	~9,600	10,440	5,300	~3,600	TBD
Delta IV M+(5,4)	~$109	~12,300	~11,200	11,800	6,120	~4,100	TBD
Delta IV Heavy	~$164	~23,400	~21,200	23,250	10,843	~7,400	TBD
Taurus (Comm/Gov)	~$20	1,250 / 1,050	1,030 / 875	1,150 / 1,090	448 / 400	320 / 290	N/C
Titan IVB	~$423	?	?	N/A	N/A	?	
*Zenit 2 (SL7)	~$45	~13,000	?	11,100	N/C	N/C	
*Zenit 3SL	~$90	N/A	N/A	N/A	5,150	3,100	see following sheet

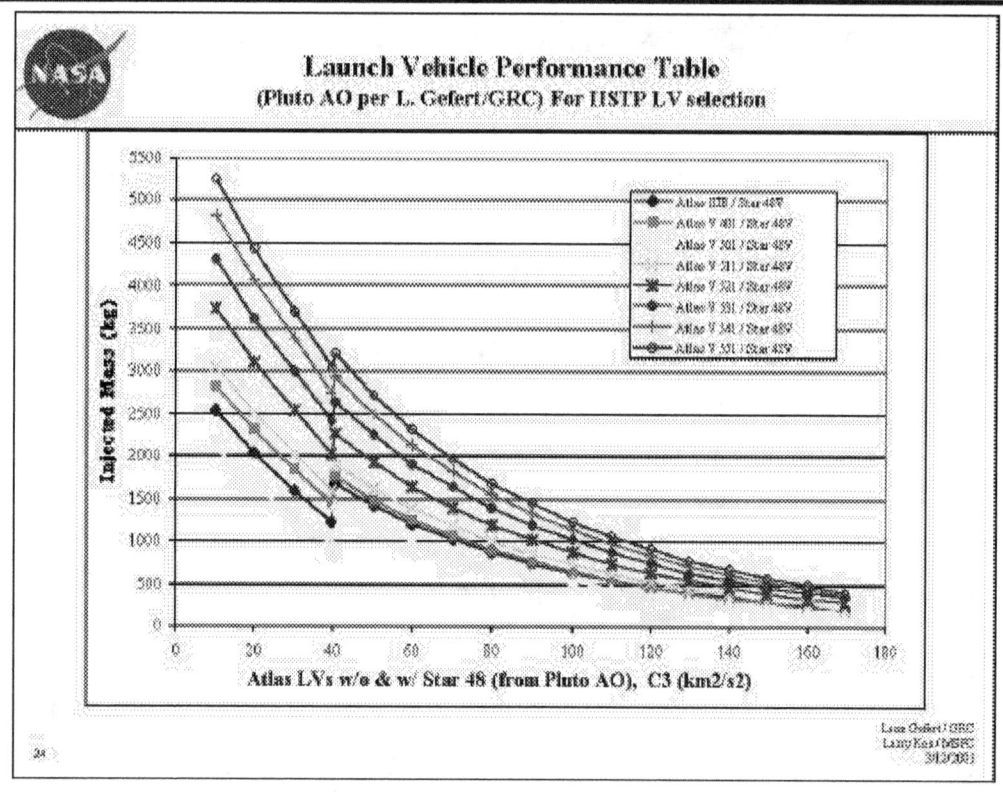

2010 Titan Organics Orbiter/Lander

Roy Kakuda
12/21/00

- Science Objectives:
  - Titan lander reference mission from the CSWG reports in priority order:
    1) Distribution and composition organics
    2) Organic chemical processes, their chemical context and energy sources
    3) Prebiological or protobiological chemistry
    4) Geological and geophysical processes and evolution
    5) Atmospheric dynamics and meteorology
    6) Seasonal variations and interactions of the atmosphere and surface (not addressed in a mission of short lifetime)

Titan Organics Orbiter/Lander Reference Mission

- **Measurement Objectives**

Instrument	Science Objective Description	CSWG Number
GCMS	Ices, evolved gases, small organics, isotopes (C,N,O), atmospheric gases	1, 2
UV/Vis/NIR spectrometer	Landing site chemical context	2, 4
NA Camera	Detailed morphology of landing site	4, 5
WA Camera	Gross morphology of landing site	4, 5
Chem lab 1	Bulk inorganic chemical properites such as redox potential, pH, electrical conductivity	2
Chem lab 2	Organics and biology	1, 3
XRFS with X-Ray Source	Chemistry of surface samples	2
Descent Radar altimeter	Surface roughness and subsurface structure to 1 km	4
Entry ASI	Atmospheric pressure, temperature and acceleration during landing	5
Sampling system	for use with GCMS, Chem lab 1 and 2, XRFS	see related instruments

Titan Organics Orbiter/Lander Reference Mission

Lander Strawman Payload

- Lander Strawman Payload
  - GCMS
  - UV/Visual/Near IR line spectrometer
  - Near Angle and Wide Angle imagers
  - 2 chemistry labs
  - XRFS
  - Entry ASI
  - Descent altimeter
  - Sample acquisition and handling system

Titan Organics Orbiter/Lander Reference Mission

Orbiter Strawman Payload

- SAR/Altimeter
- Radio Science (USO)
- Narrow Angle Imager (wave length TBD)
- Infrared radiometer

Titan Organics Orbiter/Lander Reference Mission

- Mission Design and Constraints
  - Mission Dates
    - Earliest Launch Date = 8/10/2010
    - Earliest Arrival Date = 4/17/2019
  - Launch Vehicle
    - Launch vehicle: Delta III Blue Book Cost $81-86 M FY 01 ($105M assumed)
      - Capability - Injected mass = 2208 kg @ C3 : 5.5 with no contingency
      - Shroud - 10L
  - Trip Time = 8.75 years
  - Trajectory = SEP VVGA
    - Minimum/Maximum Solar Distance - 0.64 - 10.1 AU
    - Minimum/Maximum SEP Solar Distance - 0.64 - 3.0 AU
    - Minimum/Maximum Earth Distance - 0 - 11.0 AU
  - Delta-V Requirements (post launch)
    - SEP Delta-V = 5.6 km/s
    - Chemical Delta-V = 300 to 800 m/s

Titan Organics Orbiter/Lander Reference Mission

- Mission Design and Constraints (continued)
  - Titan Approach and Tour
    - Flyby Geometry and Constraints
      - Saturn V infinity = 5.77 km/s @ DAP = -8.8 deg.
      - Titan Entry Velocity = 5.2 km/s
      - Titan exit velocity = 2.3 km/s
      - Titan aerocapture delta V = 2.9 km/s
    - Titan Orbital Parameters
      - 1400 km circular
      - 42,000 km x 1400 km initial orbit
    - Science Constraints
      - Global mapping of Titan surface, resolution TBD (limited by earth link, antenna size, radar frequency, power, etc.)
    - Orbital Maintenance (Delta-V Requirements)
      - Less than 300 m/s assumed

Titan Organics Orbiter/Lander Reference Mission

Delta 10L Launch Vehicle Envelope
Star 48

- Size of Dynamic Envelope (m)
  D = 2.74, L = 3.66, L' = 2.28,
  D' = 0.61
- Usable length "L" reduced because of Star 48 upper stage.

Titan Organics Orbiter/Lander Reference Mission

- **Mission Environment**
  - Thermal Environments
    - Landers
      - 80 °K (low atmosphere) to 180 °K (high atmosphere)
      - 90 to 97% Nitrogen and Argon, 2 to 10% Methane, 0 to 1% other hydrocarbons
    - Orbiter - Aerocapture using ballute
      - 90 to 97% Nitrogen and Argon, 2 to 10% Methane, 0 to 1% other hydrocarbons
  - Radiation
    - 50 krads with an RDM of 2 added - Defined by RTG dose geometry, solar activity, Saturn radiation, and trajectory
    - SEP (solar array, PPU, thrusters, DCIU, etc) - 3 year solar dose (0.6 to 3 AU)
  - Heliocentric Distance (max and min)
    - Minimum/Maximum Solar Distance - 0.64 to 10.1 AU
    - Minimum/Maximum SEP Solar Distance - 0.64 to 3.0 AU
  - Geocentric Distance (max and min)
    - Minimum/Maximum Earth Cruise Distance - 0 to 11 AU
    - Minimum/Maximum Earth On-Station Distance - 8 to 11 AU

Titan Organics Orbiter/Lander Reference Mission

- **Mission Environment**
  - Mission Specific
    - Flyby Velocity
      - Saturn V infinity = 5.77 km/s @ DAP = -8.8 deg,
      - Titan Entry Velocity = 5.2 km/s
      - Titan exit velocity = 2.3 km/s
    - Dust (including ring avoidance) - Very low because of Titan aerocapture at 21 Saturn radii (4 x outer Saturn rings)
    - Power Requirements
      - Lander = 82 W
      - Orbiter = 309 W
    - Data Rates
      - Lander / Orbiter link = 128 kb/s
      - Orbiter down link = 2.5 kb/s
    - Data Volume (TBD)
  - Lifetime - 8.75 year cruise, 3 year Titan orbit

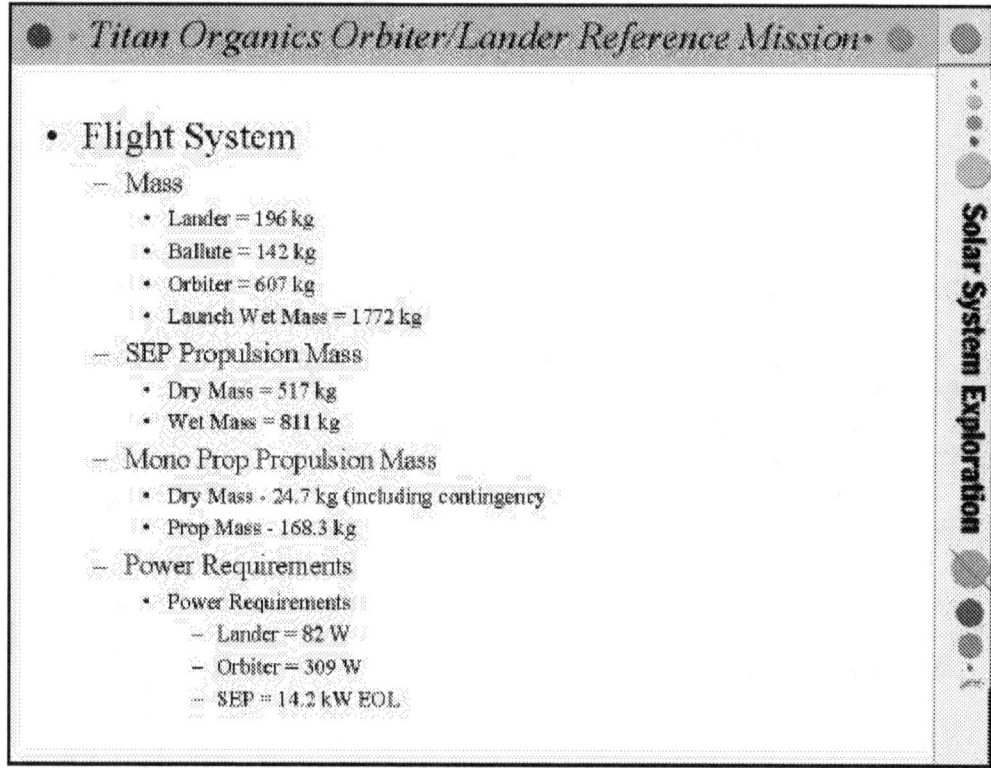

Titan Organics Orbiter/Lander Reference Mission

- Flight System
  - Mass
    - Lander = 196 kg
    - Ballute = 142 kg
    - Orbiter = 607 kg
    - Launch Wet Mass = 1772 kg
  - SEP Propulsion Mass
    - Dry Mass = 517 kg
    - Wet Mass = 811 kg
  - Mono Prop Propulsion Mass
    - Dry Mass - 24.7 kg (including contingency)
    - Prop Mass - 168.3 kg
  - Power Requirements
    - Power Requirements
      - Lander = 82 W
      - Orbiter = 309 W
      - SEP = 14.2 kW EOL

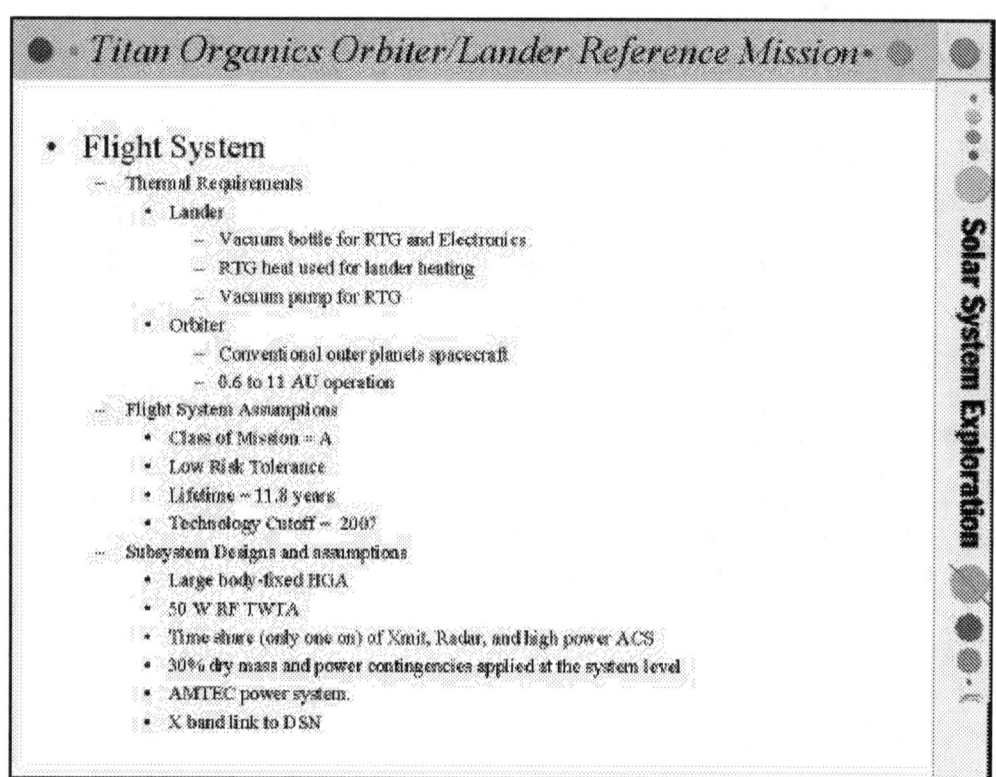

Titan Organics Orbiter/Lander Reference Mission

- Flight System
  - Thermal Requirements
    - Lander
      - Vacuum bottle for RTG and Electronics
      - RTG heat used for lander heating
      - Vacuum pump for RTG
    - Orbiter
      - Conventional outer planets spacecraft
      - 0.6 to 11 AU operation
  - Flight System Assumptions
    - Class of Mission = A
    - Low Risk Tolerance
    - Lifetime ~ 11.8 years
    - Technology Cutoff – 2002
  - Subsystem Designs and assumptions
    - Large body-fixed HGA
    - 50 W RF TWTA
    - Time share (only one on) of Xmit, Radar, and high power ACS
    - 30% dry mass and power contingencies applied at the system level
    - AMTEC power system.
    - X band link to DSN

Titan Organics Orbiter/Lander Reference Mission

- **Ground Systems and Mission Operations**
  - Data Volume
    - Lander - up to 1.3 Gbits per week
    - Orbiter - up to 1 Gbits per day
  - Data Return Strategy at Titan
    - 128 kbps from lander to orbiter
    - Five 35 minute passes (within one day) every 7 days
    - 25 kbps from orbiter to earth using 70 M passes
    - 7 8-hour 70 M passes per week, 14 8-hour 34 M passes per week
  - MOS Assumptions
    - Cruise - 8.75 years
    - Titan science operations - 3 years
  - Operations Concept
    - JPL Operations
    - Low staffing for cruise
    - Staff up at Titan minus one year

Titan Organics Orbiter/Lander Reference Mission

- **Programmatic Results**
  - Total Mission Cost - 1010 M$ in FY 2001 $
    - Orbiter - 216 M$
    - Lander - 132 M$ (including 3 M$ for the ballute.
    - SEP - 30 M$ (Customer ROM - not a Team X provided cost)
    - Assumes no technology development is paid for by the project.
  - Schedule
    - Phase A start - August 2006
    - Phase A duration - 3 months
    - Phase B duration - 9 months
    - Phase C/D duration - 36 months
    - Phase E duration - 144 months
      - Cruise - 108 months
      - Science ops - 36 months

Titan Organics Orbiter/Lander Reference Mission

- **Programmatic Constraints**
  - Orbiter and lander required
  - Post Cassini science
  - Class A mission

Titan Organics Orbiter/Lander Reference Mission

- **Liens and Open Issues**
  - Environmental Uncertainties
    - Unknown Titan Surface causes landing method and outcome uncertainty
    - Unknown Titan Surface winds causes landing method and outcome uncertainty
    - Unknown Titan atmosphere causes aerocapture method and outcome uncertainty
  - Technology development
    - The AMTEC RPS, ballute, and X2000 avionics are developed by others
    - Flight test of ballute before 2007
    - Test facility for lander / RTG (simulated Titan atmosphere) is no cost
    - Development of RTG for Titan atmosphere is no cost
  - Launch Opportunities
    - Ballistic
      - Direct: Almost every year
      - JGA: 2015 (Low Jupiter flyby), 2016, and 2018
      - VVGA: 2010, TBD
      - VVJGA: TBD
    - SEP
      - JGA: TBD
      - VVGA: 2010, TBD
      - VVJGA: TBD

IISTP Phase I Final Report
September 14, 2001

APPENDIX E

Technology Assessments

IISTP Phase I Final Report
September 14, 2001

NEP Independent Assessment
Gordon Woodcock
Gray Research, Inc.

1.0 Introduction and Purpose

This assessment was conducted as part of Gray Research support to the MSFC In-Space Integrated Space Transportation Planning activity. The purpose of the assessment was to develop an independent understanding of the performance potential for nuclear electric propulsion, and the technology characteristics that would best serve utilization of nuclear electric propulsion for future space missions.

2.0 Nuclear Electric Propulsion Mission Considerations

<u>Basic Principles</u> - Electric propulsion systems are power-limited, in contrast to chemical propulsion systems, which are energy-limited. By power-limited we mean that system design is dominated by consideration of the fixed mass of hardware needed to generate the necessary power. Energy-limited systems design is dominated by the mass of propellant needed to produce the mission energy.

Ideal velocity increments (delta Vs) for in-space transportation missions range from a few to over 20 km/sec, with most interest for application of nuclear electric propulsion falling in the range 10 km/sec to 20 km/sec. These values are large compared to the maximum practically attainable jet velocity for chemical propulsion systems, about 4.7 km/s. Achieving jet velocities for chemical propulsion as near as possible to the maximum is therefore very important, and even then, high propellant fractions and often staging are necessary. Mission designs often make use of gravity assists to enhance performance; for example, the Cassini mission to Saturn used four such assists. The large propellant mass required to achieve high propellant fraction increases the launch mass required, and places great premium on minimizing spacecraft mass. Both effects are costly.

Electric propulsion can achieve any desired jet velocity, up to the speed of light (3×10^8 m/s). However, the mass required to produce the jet is a limiting factor, and this leads to an optimum Isp for any mission, depending on mission parameters and the performance of the electric propulsion system. Consider what is required to accelerate a 1-t. spacecraft by 20 km/s with a speed-of-light jet. The momentum transferred is 20 million kg-m/s = 20 million N-s. The momentum of light is E/c where c is the speed of light. The energy required is $(20 \times 10^6)(3 \times 10^8)$, = 6×10^{15} Joules = 1670 GWh, the output of a 1000-megawatt electric powerplant for about 2½ months. We must convert 0.06 kg of mass to radiation energy. With nuclear fission, considering typical powerplant efficiency, about 200 kg of uranium must be fissioned to generate this much energy.

If, however, we use a jet velocity 40 km/s (roughly optimum) the mass ratio is 1.65 and, neglecting electric propulsion mass, the propellant required is 650 kg. The energy required to accelerate the propellant is 5×10^{11} Joules, over 2½ months, 80 kW. This is a typical power output for a near-term space nuclear powerplant. On the other hand, if chemical propulsion were used to deliver the 20 km/sec the propellant required would be, again neglecting the mass of the propulsion system, about 70,000 kg. It is clear from this example that we need "enough" jet velocity but more jet velocity is not always better.

Options - Nuclear power is one of the main options for electric propulsion, the other being solar power. Beamed power, e.g. from a laser or microwave power beaming station on Earth, has also been investigated, and isotope power has been proposed. Nuclear power has the obvious advantage that its power availability does not depend on distance from the Sun. Some missions need power and/or propulsion far from the Sun, and nuclear power is the clear choice (for power levels of watts to hundreds of watts this may mean isotope nuclear power). At high power levels (multi-hundred kilowatts and up) it appears to offer mass advantages over solar power. On the other hand, at power levels below 100 kWe, solar power has the mass advantage. Solar electric systems also have a lifetime advantage for most applications, but either system offers lifetimes on the order of years.

Mission Applications - Table 2-1 presents a summary of mission types reviewed by IISTP, and expected constraints and applications for nuclear electric propulsion.

Table 2-1: Potential Applications for Nuclear Electric Propulsion

Mission Type	Expected Application/Utility
Inner solar system or Earth vicinity	Costs and environmental risks probably exceed benefits
Outer solar system complex (robotic)	Highly applicable; unique capability to generate high-performance electric propulsion far from Sun. ~ 100 kWe
Beyond solar system	Expect reasonable capability to deliver ~ 20 year trip to ~200 a.u. Operating times may be long. Unique capability to generate high-performance electric propulsion far from Sun. ~100 to 500 kWe
HEDS lunar	Costs and environmental risks probably exceed benefits; requires high power ~ 1 MWe
HEDS Mars/Asteroid	Requires high power ~ 10 MWe. May offer fast trips at very high power ~ 50 MWe and low specific mass < 5 kg/kWe. There is a reactor disposal issue (see below), and an issue with operation in Earth orbit. Weak Stability Boundary gateway basing may be appropriate. Very high power systems expected to be expensive to develop and operate.

3.0 Issues

3.1 Safety and Integration Factors

- Public safety constraints for nuclear electric propulsion have not been defined. Their definition will be controversial. Expected constraints are as follows:

- No sustained operation will be permitted in any Earth orbit. Reactor operations in Earth vicinity create problems for gamma ray astronomy, and if at low altitude generate carbon-14 by neutron capture in the atmosphere. While for small reactors this effect may be negligible, it will be a source of controversy.

- Return to Earth orbit will not be permitted because of concern over inadvertent reentry of the reactor.

- Launch of reactors will be limited to zero-power reactors with negligible fission product inventory (thus, negligible radiation hazard in event of a launch accident).

- Reactors will be designed or equipped to remain subcritical on water immersion, i.e. in event of a launch accident.

- Testing (on Earth) requires containment/decontamination of reactor under test for normal operation as well as of loss-of-coolant accident. Facilities exist for safe testing up to multi-hundred-kilowatt thermal power levels.

In addition, there are certain integration issues:

- There has been a long-standing controversy over whether a test of a complete power generation system is required, or whether the reactor can be tested separately from the power conversion system.

- Life needs to be 1 to 2 years; life testing will be required.

- Protection of payloads and/or crew requires shielding. The extended thermal radiator geometry creates a potential backscatter source not present for an NTP reactor. Vehicle geometry can be arranged such that the radiator is shielded by a shadow shield, eliminating the backscatter problem.

- Spent reactors need to be disposed of properly, i.e. not on trajectories which could experience future Earth encounter.

- Minimum reactor size for criticality leads to a minimum practical power level presumably about 100 kWth ~ 20 to 40 kWe.

3.2 Ranges of Achievable Mass/Power Performance

To provide an indication of the useful range of mass/power performance, the following example is offered: Calculations are normalized to a unit mass (1-kg) spacecraft, which is presumed to be 75% powerplant and propulsion and 25% customer payload. Propellant is added to the 1 kg.

Propellant Mass = 0.65*burnout mass = 0.65 kg
Burnout mass = 75% powerplant & propulsion
Jet velocity, V_j = 40 km/s
Jet power = $mV^2/2$ = 8×10^8 watts for 1 kg/sec mass flow
For typical efficiency, electric power ~ 13×10^5 kWe for 1 kg/sec
Powerplant & propulsion = 0.75 kg/kWe = 0.075 kWe
Flow, kg/sec = $0.075/13 \times 10^5$ = 5.7×10^{-8}
Duration = $0.65 \text{kg}/5.7 \times 10^{-8}$ kg/s = 132 days

For most missions, the velocity needs to be delivered in less than 2 years as a maximum. Multiply 10 kg/kWe by 730/132 to obtain 55 kg/kWe as a rough maximum acceptable mass/ power ratio. For human missions to Mars and return, on opposition-like profiles (i.e. fast round trips) the calculated power duration of 132 days is already about as long as we would wish to entertain, so for these missions, the mass/power ratio should be less than 10 kg/kWe.

Many studies and papers have been published on mass/power performance for nuclear electric propulsion systems. Reasonable agreement seems to exist for near-term technology, 100 kWe-class systems. Near term technology typically implies uranium oxide/stainless steel heat-pipe-cooled reactor technology, Brayton cycle energy conversion, and rotating electromagnetic generation of electricity. At lower power levels, Stirling cycle energy conversion may offer better mass/power performance. Several energy generation cycles have been proposed and analyzed, as summarized in Table 4-1 in Section 4.

Mid-term technology is usually considered to employ refractory metal reactor fuel elements, probably still with uranium oxide, and heat pipe cooling. Turbines may require refractory materials, but the heat exchangers, except for the heat pipe unit, could be made of conventional materials.

Advanced technology implies direct reactor cooling by the cycle gas flow, graphite or carbide reactor fuel elements, and advanced materials for turbines and the recuperator heat exchanger. Note that a substantial technology legacy exists from the "high-temperature gas-cooled reactor (HTGR)" commercial power reactor programs in the UK and Canada.

3.3 Specific Observations Regarding Performance Estimates

<u>Turbine temperatures:</u> For helium gas-cooled reactors and turbines, it should be possible to use high-temperature materials which are not usable in chemically reactive gas flows. Carbon-carbon or carbon-SiC blades should be serviceable in a helium environment and could operate at temperatures above those considered practical for jet engine turbines, which operate in a hot oxidizing environment.

<u>Reactor temperatures:</u> Some authors seem to have extrapolated from nuclear rocket reactor experience, which has demonstrated 1-hour life and hoped for 10-hour life, to 10,000 hour life at the same reactor operating temperature. This is a major extrapolation. As far as I know, there is no test experience with graphite-based core materials at such lifetimes. The life limit in the nuclear rocket environment is hydrogen corrosion, which does not apply to an inert-gas-cooled reactor. However, fission products and fission product gas release, radiation damage, as well as other degradations, are applicable to long-life reactors and were not considered in the nuclear rocket case because life was limited due to hydrogen corrosion. If the helium flow is seeded by cesium (for an MHD generator), reactions between cesium and the hot reactor core must be evaluated and may affect temperature limits. Cesium has one stable isotope, which has a neutron cross section low enough to not be concerned about poisoning the reaction, but high enough to be concerned about depleting the seed concentration.

My view is that temperature limits in the range 1500K - 2000K should be considered as more realistic. There is a lot of operating experience with high-temperature gas-cooled reactors for commercial power generation. These were also graphite, helium-cooled. Maximum short-term fuel temperature (hot channel max) was cited at about 1600K, with normal fuel operating temperature about 1150K. Fuel was rated at 3 full-power years, with burnup approaching 100,000 MWD/t. (Another source gave 50,000 MWD/t.) Note that these reactors used a highly enriched U235 load, with thorium 232 as a "phoenix fuel" rather than U238.

<u>Reactor:</u> For this application, the reactor design must include burnup as well as heat transfer limits. Rocket reactors have very low burnup and it is not an issue. They are also high pressure drop designs; closed-cycle Brayton systems must be very low pressure drop. See analysis to follow.

<u>Superconducting Magnets:</u> The referenced paper describes superconducting magnets for producing the magnetic field for the MHD generator. These are presumably located near the reactor. The reactor will leak a megawatt or so of radiation ... neutrons and gamma rays. Some of this (a kilowatt?) will be deposited in the magnets. Removing heat from a superconducting magnet at liquid helium temperatures is a prodigious task. There is a tradeoff among distance from the reactor, shielding and cryostat mass, to minimize total mass penalty. We can be confident this mass penalty is greater than zero.

Turbo-compressors: Specific mass projections, based on aircraft engine experience, appear to be applicable. Note that a helium compressor may be considerably more massive. Air has 7 times the molecular weight of helium, and hence 7 times the density and 40% the speed of sound. A helium compressor is likely to need at least twice the number of stages for a given pressure ratio compared to an air compressor. Some analysts have proposed helium-xenon mixtures to solve the molecular weight issue; the mix apparently has most of the conductivity and heat capacity per unit volume of helium but is much easier to pump.

In an MHD design, an electric motor must be used to drive the compressor, and appears to have been neglected in some references. Its specific mass will be many times that of the compressor. I referred back to one of the solar power satellite thermal cycle studies of several years ago. It described a 32-megawatt electrical generator at 0.14 kg/kWe, not including its thermal control system. This estimate was made by General Electric, a builder of high-power aerospace electric generators.

Of course, if one uses a conventional turbine, the compressor may be driven by a shaft but the power output must come from a generator which will be as heavy per unit power as the motor. Note that for a typical closed Brayton cycle the compressor power is about twice the output power, so the advantage still goes to the conventional turbine.

Regenerator (also called recuperator): The regenerator mass per unit heat transfer area is estimated as 1 kg/m^2. This may be appropriate for a lightweight, moderate-temperature industrial design. Note that if the recuperator is a tube-in-shell design, the mass of a tube is pDLtp (thin wall approximation) where terms are D diameter, L length, t thickness, and ρ material density. The heat transfer area is pDL, and the ratio m/A is just tρ, which is intuitive. For the temperatures of operation, up to over 1400K (over 2100F) the material must be a turbine-type nickel-based alloy. For these, ρ is about 8000 kg/m^3. For m/A to be 1 (just for the tubes), the wall thickness must be 0.125 mm = 0.005".

Radiator: The radiator mass per unit area is a significant contributor to overall mass. 1 kg/m^2 is equivalent to a sheet of aluminum 1/2800 m = 0.36 mm thick. This is 0.014". If the material were a copper alloy as probably necessary at the planned radiator temperatures 500 - 700K (440 - 800F), the thickness would be 1/8000 = 0.125 mm = 0.005". Small fin radiators on spacecraft may indeed be so thin, but this radiator is another animal entirely and will be several times as massive. One cannot afford the mass penalty, pressure drop, or leak risk of piping the helium all over the large radiator area (for the cycle I analyzed, 10 MWe, the radiator area is about half a football field). Therefore, the design needs to be a compact(!) heat pipe heat exchanger which transfers waste heat from the helium flow to a large number of heat pipes which then distribute the heat over the radiator area. It will be > 1 kg/m^2.

MHD vs turbine: As cycle peak temperatures are reduced in the interest of realism, and radiator masses become more realistic, the higher efficiency of a turbine versus an MHD generator, combined with the reduced size of output generator versus compressor drive motor, may tip the balance in favor of a conventional turbine, if turbine materials and designs can be developed for helium use at selected cycle temperatures. The tradeoff should be based on point designs for comparative systems at realistic temperatures and component mass characteristics.

3.4 Sensitivities

Mission/performance sensitivities and representative estimates are presented in Figure 3-1.

- Specific power is sensitive to technology level and power output.
- NEP does not scale to low power well.
- May not make sense to produce a reactor at less than 100 kWe capability.
- Thruster sensitivities will be the same as for SEP. High values of system specific power favor high efficiency thrusters, e.g. ion. Low values favor low mass thrusters, e.g. MPD.
- Thruster selection is also driven by mission optimum Isp. Some NEP missions need high Isp > 5000 sec, for which ion thrusters may be the only practical solution. At high power, some of the plasma devices may work well.

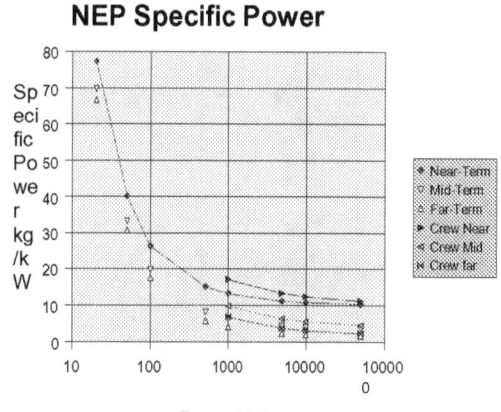

These projections were supplied by Bob Cataldo of the NASA Glenn Research Center. Where overlap in estimating existed, we generally agree.

Figure 3-1: Performance Sensitivities - Discussion

Estimates from other sources, especially at high power levels, varied widely, with some estimates well below 1 kg/kWe. Some of these estimates were linked to MHD generators (rather than turbine-generators). Others considered gas-phase (plasma) reactors along with MHD.

Since the efficacy of nuclear electric propulsion for human Mars missions seems to depend on achieving low values of mass/power, the present investigation was focused on high-power advanced technology reactors.

4.0 Selection of Systems for Analysis

A brief review of potential cycles was performed, as summarized in **Table 4-1.**

Potential Cycles	Considerations
Thermoelectric	Cycle efficiency very low and max temperature restricted; thus mass/power relatively high.
Thermionic	Promise of good efficiency has never materialized; plagued by materials problems.
Brayton	Tends to large radiator areas but cycle is high efficiency.
Turbine	"Traditional" design; turbine temperatures may be limiting.
MHD	Potential for high cycle temperatures if reactor materials and life are capable.
MHD gas-core	Removes reactor (but not other) temperature limits; very speculative and difficult to develop.
Rankine	Higher average radiator temperature for same cycle bottom temperature; working fluids usually corrosive.
Steam	Classical terrestrial thermal power cycle; radiator temperatures too low for space.
Liquid Metal	SNAP-8 tried mercury (nasty material); modern designs use potassium; materials problems rampant.
Stirling	Because it involves a lot of heat exchange, tends to be preferred only for low-power (10's kW) systems.

Based on the considerations in the table, Brayton turbine and MHD cycles were selected. A specific objective was to estimate the advantages for MHD generation.

5.0 Cycle Analysis

The specific cycle analyzed was taken from the referenced paper. It is diagrammed in Figure 5-1. Helium is compressed by a compressor, shaft-driven in the case of a turbine expander and motor-driven in the case of an MHD expander. Two intercooler stages are used to reduce the average heat rejection temperature. This improves cycle efficiency for a given cycle temperature ratio, but increases the radiator area per unit heat rejection. There is an obvious trade here; the trade was not performed.

Helium leaves the compressor and enters a recuperator which preheats it by transferring heat from the helium leaving the turbine or MHD expander. This also improves cycle efficiency by increasing the average cycle temperature ratio for a given max/min temperature ratio. The recuperator enables practical cycle efficiencies above 25%, not otherwise achievable.

Leaving the recuperator, the helium enters the reactor where it is heated to the cycle maximum temperature. It then enters the expander (MHD or turbine). Leaving the expander, the helium enters the recuperator where it is further cooled by transferring heat to the compressor discharge flow. Leaving the recuperator the helium enters the radiator heat exchanger and is cooled to the cycle minimum temperature.

State points are presented in the Figure. Red text shows a representative MHD expander case, with maximum temperature 2000K, and black data are for a turbine expander with maximum temperature 1500K. These values represent my estimates of maximum practical cycle temperatures for these cases. Cycle minimum temperature was not optimized but is not far off optimum. Temperatures are K and mass flows kg/s.

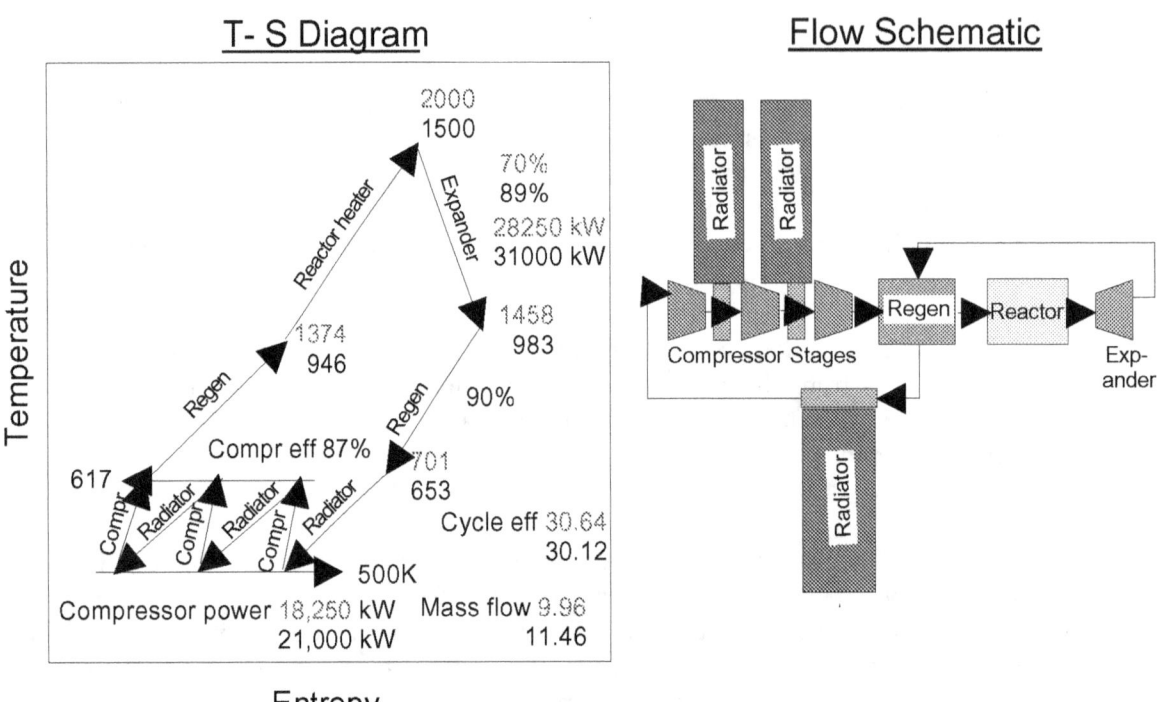

Figure 5-1: Brayton Cycle Diagram

Pressure Drop Effect on Cycle Efficiency: We used the same cycle diagram as the referenced paper. Cycle points are as follows:

1 - Compressor 1st stage inlet
2 - Compressor 1st stage exit
3 - Compressor 2nd stage inlet
4 - Compressor 2nd stage exit
5 - Compressor 3rd stage inlet
6 - Compressor exit/recuperator inlet (cool side)
7 - Recuperator exit/reactor inlet
8 - Reactor exit/expander inlet
9 - Expander exit/recuperator inlet (hot side)
10 - Recuperator exit/radiator inlet

The pressure ratio across the expander can be expressed as

P8/P9 = P8/P6*P6/P1*P1/P9
      = P8/P7*P7/P6*P6/P5*P5/P4*P4/P3*P3/P2*P2/P1*P1/P10*P10/P9

and noting P6/P5*P4/P3*P2/P1 = r_c^N,

P8/P9 = P8/P7*P7/P6*P5/P4*P3/P2*P1/P10*P10/P9* r_c^N

where all the pressure ratios on the right hand side of the latter expression are pressure drop ratios, which can be combined, to express

P8/P9 = G* r_c^N, where G <1 is the net pressure drop ratio for the entire cycle.

Using the authors' expression for cycle efficiency,

$$\eta^{th} = \frac{T_{max}(1 - \frac{T_9}{T_{max}}) - N_c T_{min}(\frac{T_2}{T_{min}} - 1)}{T_{max}(1 - \frac{T_7}{T_{max}})}$$

We determine T9/Tmax as $1 - h_{s,g}[1 - 1/(Gp_g)^{(\gamma-1)/\gamma}]$;

T2/Tmin as $1 + 1/h_{s,g}[p_g^{(\gamma-1)/(N\gamma)} - 1]$;

E-15

and   T7/Tmax as er (T9/Tmax) + (1-er)(T6/Tmax)

By assumption, T2 = T6; therefore (T2/Tmin)(Tmin/Tmax) may be substituted for T6/Tmax. Using these expressions, one can plot cycle efficiency versus cycle pressure ratio for a range of values of pressure drop ratio, as is done in Figure 5-2.

For purpose of analysis of achievable power-to-mass ratio, I selected the top center chart with pressure ratio 4 and pressure drop ratio 0.85, and cycle efficiency 30%. This reflects my skepticism of operating the reactor with a helium outlet temperature of 2500K for a long period of time. The pressure ratio is near optimum; I saw no reason to stay with the reference pressure ratio 8.

I also analyzed a representative turbomachine (as opposed to MHD) conversion cycle, with cycle maximum temperature 1500K and minimum temperature 500K, also with pressure drop ratio 0.85. This case, coincidentally, also has cycle efficiency 30%.
Full optimization of the cycle requires optimizing on pressure ratio, low temperature limit (assuming high temperature is fixed at maximum hardware capability), pressure drop versus mass of each major component, and radiator design.

I used a small C code to generate the cycle efficiency curves and a spread sheet to analyze mass/power ratio. Cycle state points were picked off from the C code and manually transferred to the spread sheet.

Reactor Performance: The reactor design was assumed cylindrical, similar to a NERVA reactor. Two considerations were used to size the reactor: fuel burnup and heat transfer. For simplicity I assumed the reactor core was $U^{235}C_2$ and graphite. A practical design might add thorium-232, as needed to get the right criticality and to provide some breeding to counteract burnup. No neutronics analyses were done. The reactor is certainly large enough. The main reasons for a neutronics analysis are to size the reflector, assess controllability based on reflector drums, and determine reasonable burnup and benefits of thorium addition.

Cycle Parametric Analysis Results

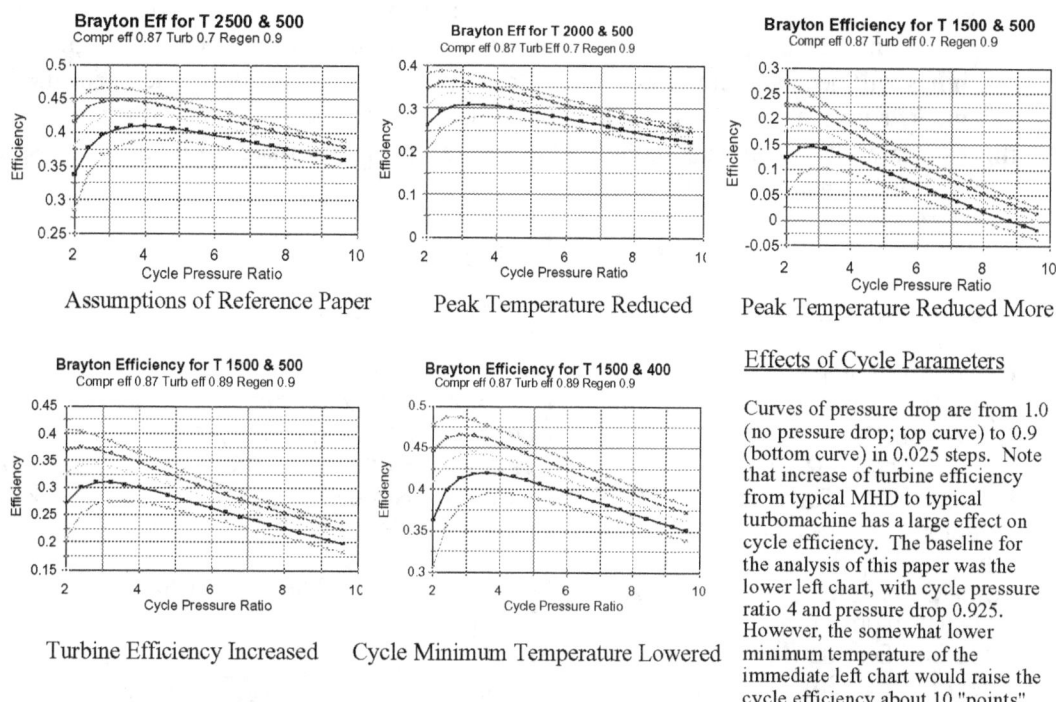

Figure 5-2

Fuel load was based on 80,000 MWD/ton, about 9% burnup, and the physical size of the reactor was based on a 20% void fraction for helium passages, an assigned pressure drop of 3 psi (about a fifth of the allowable for the entire circuit), and the necessary heat transfer area. The graphite mass was determined by balance of volume after fuel load. Viscosity was determined by a kinetic theory relationship:

$m = 2.6693 \times 10^{-5}$ sqrt $(MT)/(d^2 \Omega)$ where the result is in cgs units. For mks units, divide by 10, which was done on the spread sheet.

Averages were used, where a real heat transfer analysis would consider several points in the helium passages to assess heat transfer versus helium temperature and other flow conditions. The Reynolds' number in the passages is lower than I would like, but is probably OK. Friction coefficient was an assumed value. A 20 cm (8") reflector was assumed, with reflector controls assumed included in the reflector mass. The reactor size result is somewhat too small for mass flow (ρAV), so further design iteration would be required for a real design. However, this seems to be in the ballpark. Main reactor parameters are given in Figure 5-3.

Turbomachine: Used a specific mass of 0.025 kg/kW shaft power. Various sources suggest this is about right. However, none of these sources described helium turbomachines; it is quite possible that because of the low molecular weight, helium machines will need so many more stages they will be significantly heavier. For the MHD expander, I used a specific mass of 0.05 kg/kWe. There is little data on which to base this estimate. It has only a small effect on overall power-to-mass ratio unless the specific mass is much greater.

	MHD	Turbine		MHD	Turbine
Electric Power Output	10 MW	(same)	Heat Transfer Passage L/D	400	(same)
Thermal Power MW	32.6	33.2	Passage Size	5 mm	(same)
Cycle Max Temp	2000K	(same)	Delta P	3 psi	(same)
Cycle Min Temp	500K	(same)	Reynolds' No.	2900	(same)
Cycle Max Pressure	10 atm	(same)	h, kcal/m2-K	0.19	(same)
Cycle Pressure Ratio	4	(same)	Reactor Vol m3	2.04	2.07
Pressure Drop Ratio	0.85	(same)	Reactor Length	2 m	(same)
Reactor Void Fraction	20%	(same)	Reactor Diam	1.14 m	1.15 m
Design Life (yr)	2	(same)			
Total Megawatt Days	23,840	24,255			
Total Uranium Burn	27 kg	27.5 kg			
Assumed Burnup (MWD/t)	80,000	(same)			
Fuel Load (U235)	298 kg	303 kg			
Burnup %	9.1	9.1			
UC2 Load	328 kg	334 kg			
Graphite Mass	2560 kg	2608 kg			
Reflector Thickness	0.2 m	(same)			
Reflector Mass	2148 kg	2171 kg			
Vessel Mass	540 kg	545 kg			
Total Mass & Alpha	5580 kg	5658 kg			
kg/kWe	0.56	0.57			

Figure 5-3: Reactor Parameters From Spread Sheet

Regenerator/recuperator: A tube-in-shell design was assumed, and heat transfer area required was factored from the reactor heat transfer analysis, considering delta Ts and total heat transfer required. I used a somewhat greater mass/area than in the reference paper, because the latter results in very thin wall tubes. Also, I added a calculated allowance for shell mass. Since this shell will run quite hot, I used a low stress value for the shell, and assumed it would have the density of a turbine alloy.

Radiator: Radiator area was calculated based on total heat rejection and assumed average temperature. The average temperature will trend close to or below the cycle minimum temperature because of temperature drops between the helium minimum temperature and the actual heat rejection temperature. The radiator was assumed to be a finned heat pipe design, with flat fins between the pipes externally and circular fins inside the helium-to-heat-pipe heat exchanger manifold. Sodium or potassium appear to be suitable heat pipe fluids for the temperature range considered. At a somewhat lower cycle minimum temperature, water could work. Thermal power per heat pipe, and length of the pipes, is probably pushing the state of the art. Capillary-pumped loops might be better.

I used a numerical integration to roughly iterate on fin thickness. Fins too thin, too much delta T and radiator weight goes up. Fins too thick, fins weigh too much. There is an optimum, and getting the complete optimization is a fair amount of work; for example, it also involves varying the heat pipe size and spacing. My optimization was rough, but I think the radiator mass is representative.

The radiator is actually in 3 parts. One section rejects heat in cooling the helium from regenerator outlet to compressor inlet, and the other two sections reject heat from the compressor intercooler segments of the cycle. The radiator total area is so large as to dwarf the rest of the system, although at 3743 sq m (about 3/4 of a football field) this area would only generate a little over 1 megawatt as a high-performance solar array.

Main recuperator and radiator parameters are shown in Figure 5-4.

	MHD	Turbine		MHD	Turbine
Recuperator heat transfer kcal/s	9427	4726	Radiator delta T rec out compr in	201	153
	(39,460 kW)	(19,782 kW)	Heat radiated kWth	10,468	9200
Req'd heat transfer area m2	586	675	Radiator delta T intercoolers	116.72	(same)
Tube diameter mm	6	(same)	Heat radiated (x2) kWth	6083	7001
Vol/Area m3/m2	0.0045	(same)	Estimated radiator HX area m2	1132	1160
Recuperator volume m3	2.64	3.04	Heat pipe diam cm & length m	5; 20	(same)
Total tube flow area m2	0.31	0.356	Heat pipe spacing cm	15	(same)
Number of tubes	10,940	12,593	Area per pipe m2	3	(same)
Recuperator cross-section m2	0.93	1.07	Number of pipes	1248	1279
Recuperator diameter m	1.09	1.17	Thermal power per pipe kWth	18.14	(same)
Recuperator length m	2.84	(same)	Pipe wall mm	0.2	(same)
Tube mass per unit area kg/m2	1.5	(same)	Mass per pipe kg	5.03	(same)
Tube wall thickness mm	0.2	(same)	Fin thick mm	0.2	(same)
Shell stress ksi	5	(same)	Fin area m2	2495	2558
Shell wall mm	4	4.2	Fin mass kg	3992	4092
Tube mass kg	880	1012	Radiator mass (not incl manifold)	10,263	10,520
Shell mass kg	365	425	Heat transfer area per pipe m2	0.91	(same)
Baffles & misc. mass kg	73	85	Manifold wall mm	1	(same)
Total recuperator mass kg	1318	1521	Manifold mass kg	5763	5908
Heat rejected kWth	22,634	23,203			
Radiator HTX DT	25K	(same)			
Fin DT	50K	(same)			
Average Temp	475K	(same)			
Emissivity	0.9	(same)			
Sides	2	(same)			
Heat per unit area (Stef-Boltz) kW/m2	6.05	(same)			
Area required m2	3743	3837			

Figure 5-4: Recuperator and Radiator Parameters

6.0 Results

6.1 Mass/Power Performance

The specific mass summary for the system is as follows:

Total Raw Alpha

Reactor		0.55798
Generator		0.14124
Recuperator		0.13169
Compressor & Drive		0.31936
Radiator		1.02626
Radiator Manifold		0.57629
Total		2.75285
Integration Factor	25%	0.68821
Total Estimate (kg/kWe)		3.44107

Although much greater than the estimates of the reference paper, this is still a very lightweight system compared to most estimates of space nuclear-electric systems. The reasons for the high performance are high power (10 megawatts) and high cycle temperature.

Turbogenerator System

This system differs from the reference system as follows:

Cycle max temperature 1500K instead of 2000K
Expander is turbine rather than MHD device, efficiency 0.89 instead of 0.70
No motor required to drive compressor (it's shaft-driven)
Shaft-driven rotating generator required to produce electrical power

The specific mass summary for this system is as follows:

Total Raw Alpha

Reactor		0.56587
Compressor and Turbine		0.13002
Recuperator		0.15216
Generator		0.15000
Radiator		1.05204
Radiator Manifold		0.59076
Total		2.64086
Integration Factor	25%	0.66021
Total Estimate		3.30108

Comparing the two systems, the reactor, recuperator and radiator are almost identical. Cycle efficiencies are almost the same; the reduced maximum temperature for the turbogenerator system is compensated by the greater turbine efficiency compared to the MHD machine. The rotating system has slightly less mass than the MHD machine, compressor and drive. This is mainly because the generator has about half the power rating of the compressor drive.

6.2 Desirable System Characteristics

- Desirable thruster characteristics are the same as for the SEP system.

- Reactor must not go critical upon water immersion (launch safety).

- Adequate materials life margin at highest operating temperatures.

- "Leak safe", i.e. loss of cycle working fluid highly unlikely, or redundant system which can continue to operate if some working fluid lost.

- Minimize auxiliary power required to start thermal cycle (e.g. minimize spin-up power to start turbogenerator)

- Fluid loop joining not required in space (design the system so that all fluid loops are filled and checked out on the ground and only deployed in space).

- Minimize needs for auxiliary cooling loops. (Some will almost certainly be required.)

- Ability to match power generation to thrusters with minimum of power processing and control.

6.3 Design Strategies and Approaches

For small systems ~ 100 kWe, heat pipe cooled reactors work well and can be made relatively fail-safe. Materials are generally current state of the art.

Small reactor systems could have multiple independent helium heat exchange paths to provide redundancy against helium leaks. Each path would have its own turbogenerator and heat rejection system, as illustrated in Figure 6-1.

Electric propulsion systems can usually trade partial power loss for greater trip time, so this redundancy offers ability to do a degraded mission in the event of helium loss.

Larger systems may use helium flow through the reactor; this is especially true if high power-to-mass ratio is sought. These systems appear to have less redundancy potential. Use of refractory materials for reactors, heat exchangers and turbomachines (or MHD converters) are predicted to yield specific mass < 5 kg/kWe, as noted above. Reaching these performance levels will require significant materials development, especially for reactor fuels.

The heat pipe heat exchanger has four independent helium paths, each rated at 25 kWe. Loss of one helium circuit does not shut down the others. Any number from 1 to 4 may be used. Dimensions are assumed. No criticality analysis was done.

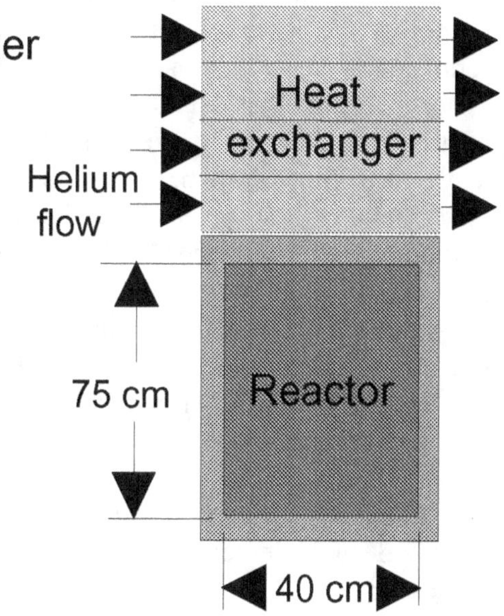

Figure 6-1: Redundant Heat Removal Concept

6.4 Risks

- Materials durability and life
- Extensive body of experience exists for certain reactor fuel forms
- Stainless steel and uranium dioxide, valid to ~1200K
- Graphite and uranium carbide in inert gas to ~ 1500K
- Graphite and various carbide-based fuels to short-term temperatures > 2500K (rocket reactor tests; carbide fuel data base is mainly non-neutronic).
- Creep-rupture criteria must be used for metallics under stress at high temperature (e.g. For reactor vessels, turbomachines, heat exchangers); generally limits nickel-based alloys to about 1250K
- Under inert gas, refractory materials can reach higher temperatures; need technology tests to set limits.
- Leakage is a major issue for long-life helium and liquid metal systems.
- The need for an acceptance test at temperature to assure no leaks conflicts with the requirement to launch inert reactor.

- Risk reduction and control should be focus of technology advancement and development plans.
- Difficult to estimate costs until this is done.

6.5 Systems Testing

Development of NEP systems is usually considered as two separate developments, one for the electric power system (including the reactor) and one for the electric propulsion system. Sometimes the division point is considered to be between the reactor/heat pipe thermal source and the generator system, which may be coupled to the propulsion hardware. The most logical break point appears to be (1) testing the reactor/heat pipe system, or in the case of a direct-cooled advanced reactor, the reactor/helium flow system, in which the power conversion equipment is simulated by a circulating pump with heat removal; (2) testing the power conversion system, initially alone and later integrated with the electric propulsion system, with electrically-heated heat pipes. This permits simulating start and stop transients with an integrated power generation and propulsion system. Electrically heated heat pipes, or an electrically-heated heat exchanger (for the direct-cooled system) can simulate the reactor up and down power ramps.

Initial testing would develop the components to the point of integrated testing. Integrated fuel element/heat pipe testing would be performed, such as currently in progress at MSFC. Nuclear component-level testing appears not required. An integrated test program would be something like the following:

(R1) Critical assembly ... neutron flux, criticality measurements, and control drum effectiveness at low or "zero" power, no coolant flow. No special facilities required. This would be a DoE test.

(R2) First reactor build: Neutron flux and heat transfer at power, with facility-pumped helium-xenon coolant mixture. This requires a reactor test facility which can contain reactor failures. Unlike the NTP, there is no potentially radioactive effluent from the reactor under normal operating conditions. Xenon has several stable isotopes, some of which have an appreciable neutron cross section (~ 5 barns). Therefore, one would expect some xenon activation to occur. Activation products all appear to have short half-lives, no more than a few days. Inadvertent leakage of activated xenon, if xenon is mixed with helium as the working fluid, may be an issue.

Tests begin with start transients, continue at reduced power and temperature, and increase in power, temperature and duration as data are collected. The test objective is to operate the reactor for the design duration. The reactor is monitored for unexpected fission product release, which would indicate a fuel element failure. At the end of the test series, the reactor is dismantled and inspected in detail in a robotic hot cell. This testing is similar to previous reactor developments. Existing facilities at one or more of the national laboratories appear adequate for these tests. If the reactor performs as expected, one build is sufficient for this test.

(R3) Second reactor build: Reactor qualification, with facility-pumped helium-xenon coolant mixture for tests at power. The clean-cold reactor is rendered safe (incapable of going critical) and subjected to launch environment tests and any other environmental tests required by the specification. It is then inspected for damage or deterioration. The reactor is then placed in a reactor test facility, probably the same one as used for the first build tests, and subjected to a qualification test including control functionality before and after a life test. The reactor is monitored for unexpected fission product release, which would indicate a fuel element failure. At the end of the test series, the reactor is dismantled and inspected in detail in a robotic hot cell.

(P1) Power conversion development: The power conversion system is fully developed by non-nuclear testing, with an electrical heat source powering the heat pipes, to simulate the reactor heat exchanger. The power conversion system is life-qualified in the same facility, including envelope excursion and life tests. Prior to life qualification, the power conversion system is subjected to launch environment and other environmental qualification tests. Existing test facilities are capable of this. Some special test equipment may be required, but is not a schedule or cost issue.

(P2) Electric propulsion development: The electric propulsion system is similarly fully developed and qualified, in an electric propulsion test facility. These facilities also exist, for the Neptune Orbiter class propulsion system.

(P3) Power integration: The power conversion system is integrated with the electric propulsion system and the assembly is subjected to start, stop and power excursion tests. Electrical heat is used to simulate the reactor heat source. These tests demonstrate simulated flight operation of the power/ propulsion system.

(R4) Third reactor build ... first flight reactor: This reactor is subjected to non-power acceptance testing and delivered for first system flight. An integrated power conversion and electric propulsion system is integrated with the reactor. Functional tests and helium leak tests are conducted on the assembled system, at the launch site. Reactor controls are exercised at "zero" power to calibrate criticality and proper control function. This final test is conducted in a shielded facility to guard against unplanned power excursions. The reactor is prepared for flight by implementing whatever launch safety provisions are specified; for example, neutron poisons may need to be installed.

In view of these requirements, the NEP program should be considered as requiring major test facilities, which apparently currently exist for reactor power levels in the 100 kWe range as appropriate for robotic missions to the outer planets. New facilities would be required for high-power reactors as appropriate for human exploration missions. Further evaluation of requirements at the launch site should be conducted for all reactor classes.

6.6 Cost Considerations

- Non-recurring:
    - These reactors, at least 100 kWe-class and current materials, do not appear expensive. Applicable data base is substantial.
    - Existing facilities may be usable, but probably require modification. A reactor containment vessel is required for nuclear testing at power.
    - Power conversion systems for 100 kWe-class systems are state of the art. Full scale development is required.
    - Systems and life testing expected to be expensive.
    - High-power, advanced-technology systems appear expensive for technology advancement and development. May need new facilities.

- Recurring:
    - Costing should be possible with CERs; help from GRC.
    - Heat exchangers are high-quality welded structures.
    - Turbomachines are similar to rocket turbopumps.
    - May want to add a little to avionics cost estimates for rad hardening.

- Operations:
    - NEP-savvy staff required for operations at least during power-up periods, which usually last for years.
    - Expect added systems safety costs to satisfy environmental safety.
    - Post-mission reactor disposal but this appears to be a minor cost.

6.7 Strategies for Technology Advancement and Mission Readiness

- Certify fuel forms and materials by thorough testing
- Small systems first to minimize costs of problems
    - NEP use on HEDS missions is more doubtful than NTP.
    - Not clear there is a foreseeable need for expensive, risky high-power systems.
- May want first in-flight use to begin at Earth escape to minimize safety and environmental issues
- Thorough ground test program in containment facility
- No reason identified for a technology flight test.
- Flight engines cannot be acceptance tested because they become radioactive. Can and should do turbogenerator "green runs".
- Long-duration qualification could use a progressive mission program

- Qualify system for 1 - 2 years by ground test
- Operate on 1 to 3 year missions
- Use flight data to increase qualified run time
- Apply to longer and longer missions

7.0 Conclusions

- **NEP is applicable to most mission categories**
    - Inner solar system complex profile
    - Outer solar system simple profile
    - Outer solar system complex profile
    - Beyond solar system
    - HEDS Mars and asteroids, but no strong advantages and has operational issues.
- **The technology is well-understood in principle**
- Reactor and power conversion technology programs.
- Mature analytical capabilities
- **Mass/power ratios less than 5 kg/kWe are probably achievable**
    - Direct-cooled closed cycle helium or helium-xenon cycle and reactor
    - Turbine-based system appears to provide performance about equal to MHD system with significantly lower maximum temperatures (e.g. 1500K vs 2000K) and more mature technology
    - Projections of mass/power 1 kg/kWe or less do not appear realistic for any foreseeable technology.
- Significant public safety and environmental issues exist
    - Operations in, and return to, Earth orbit may be restricted or prohibited.
    - New facilities appear needed for high-power systems > 100 kWe
    - Containment of accident required; may be main cost impact on test facility
    - Launch "virgin" reactors; not significantly radioactive
- Non-recurring costs require careful evaluation
- Operations costs require careful evaluation
- Loss of helium may be major risk for dynamic conversion systems.

Nuclear Thermal Propulsion (NTP) Independent Assessment
Gordon Woodcock
Gray Research, Inc.

1.0 Introduction and Purpose

This assessment was conducted as part of Gray Research support to the MSFC In-Space Integrated Space Transportation Planning activity. The purpose of the assessment was to develop an independent understanding of the performance potential for nuclear thermal propulsion, and the technology characteristics that would best serve utilization of nuclear thermal propulsion for future space missions.

2.0 Nuclear Thermal Propulsion Mission Considerations

Basic Principles - As we noted in a companion assessment, chemical propulsion systems are energy-limited. Hydrogen and oxygen, one of the most energetic reactions, and one that is practical to use, releases 57 kcal/g-mol or 57 kcal per 0.018 kg. This represents about 13.2 million J/kg, and when this is converted to kinetic energy, the ideally attainable jet velocity is 5149 m/s for an Isp of 525 sec. Actual rocket engines can convert about 80% of the energy to kinetic energy operating in vacuum, so one would expect an actual attainable Isp of 465 to 470 seconds. To do better requires an alternate energy source.

Soon after the discovery of practical release of nuclear energy in fission reactors, it was recognized that here was an enormous source of propulsion energy. The energy released in a fission reaction is about 180 Mev per U-235 nucleus. Given 1.602×10^{-13} J/Mev & 1 amu = 1.66×10^{-27} kg, we calculate the energy of fission $180 \times 1.602 \times 10^{-13}/(235 \times 1.66 \times 10^{-27}$ kg) which equals about 7.4×10^{13} J/kg and by the same logic we should be able to attain an Isp about a million seconds.

Alas, no one could figure out how to make such an engine because no imaginable material could contain such a reaction. Ordinary fission reactors dilute the reaction millions of times as the heat of the reaction is transferred to fuel elements which operate at modest temperatures. Locally, the great energy of each fission reaction creates material damage but the damage is readily dealt with by the bulk properties of the fuel element material, up to a point.

It was, however, realized that if one could operate a reactor with solid fuel elements at high temperatures and heat a light gas such as hydrogen to the fuel temperature, one might achieve Isp between 800 and 1000 seconds, about twice that of chemical rockets. While far short of the energy limit figure, this is enough improvement to be interesting. In the late 1950s, a technology program was started to exploit the possibilities. This grew into the *Rover* program which built and tested several experimental rocket reactors in the

1960s and early 1970s. These tests demonstrated achievability of the target range of Isp with operating times up to an hour at high temperature. Congress voted to terminate funding for the Rover program about 1972.

This nuclear rocket engine type exhibits the characteristics of an energy-limited system. Isp is limited to 1000 seconds or less, but high thrust is readily achieved.

Today this type of nuclear rocket is referred to as nuclear thermal propulsion (NTP).

Ideal velocity increments (delta Vs) for in-space transportation missions range from a few to over 20 km/sec, with most interest for application of nuclear thermal propulsion falling in the range 7 km/sec to 15 km/sec. Below 7 km/s chemical propulsion is relatively capable, and above 15 km/s the mass of an NTP system begins to grow rapidly.

Mission Applications - Table 2-1 presents a summary of mission types reviewed by IISTP, and expected constraints and applications for nuclear electric propulsion.

Table 2-1: Potential Applications for Nuclear Thermal Propulsion

Mission Type	Expected Application/Utility
Pluto Flyby	Delta V ~ 17 km/s from LEO or ~ 13 km/s from Earth escape; may need launch to escape, or staging.
Europa Orbiter	Trade vs. chemical + gravity assist.
Neptune or Pluto Orbiter	May need staging; Trans-Neptune/Pluto injection stage and capture stage. Can use aeroassist at Neptune. Pluto capture much greater delta V.
Mars Sample Return	Needs reactor disposal strategies for TMI & TEI. TMI disposal may merely require gravity assist at Mars
Kuiper Object Rendezvous	Similar to Pluto Orbiter
Interstellar Probe (to 200 AU)	Staged case may be interesting: Launch to Earth escape & then 2 NTP stages to add ~ 25 km/s
HEDS lunar	Needs reactor disposal strategy; benefit vs cost?
HEDS Mars/Asteroid	NTP is one of the reference systems. Needs reactor disposal strategy.

3.0 Issues

3.1 <u>Safety and Integration Factors</u>

Public safety constraints for nuclear electric propulsion have not been defined. Their definition will be controversial. Expected constraints are as follows:

Probably no sustained operation in any Earth orbit, but a burn leaving Earth orbit probably OK. Start altitude may be higher than for chemical stage.

Probably no return to Earth orbit.

Launch clean reactors (thus, negligible radiation hazard on launch accident).

Reactors designed or equipped to remain subcritical on water immersion (launch accident protection).

Testing (on Earth) requires containment/decontamination of exhaust as well as of loss-of-coolant accident.

Afterheat, or jettison of engine or entire stage, removal required following high-power burn.

Protection of payloads and/or crew requires shadow shielding.

Life limited to 1 - few hours (normally not a practical limitation).

Spent reactor needs to be disposed of properly, i.e. not on a trajectory which could experience future Earth encounter.

Hydrogen propellant for high Isp (operating temperature limits)

Minimum reactor size for criticality leads to a minimum practical thrust level ~ 2K

3.2 Ranges of Achievable Mass/Power Performance

An estimate for a small NTP was prepared using a spread sheet. Results are shown in Figure 3-1.

- Thrust-to-weight is dictated by reactor size, reflector, controls, shielding, and installation masses.
- Representative calculations are shown to the right.
- The break in the curve below occurs because the reactor is assumed to require a minimum of 2 ft³
- Integration mass is that required to make a reactor and pump into an operating engine, i.e. valves, ducting, structure, and controls.
- Installation mass includes thrust structure and gimbal provisions.

Engine Characteristics	
Assume Thrust	3000 lbf
Assume Temp	2700 K
Mol Wt	2.016
C* num	12481.343
C* den	0.761
C*	16405.634 ft/s
Isp	892.872 sec
Mass Flow	3.36 lb/s
Thermal Power	14402.378 kcal/s
	60.288 MWth

Estimate of Engine Mass	
Reactor Power Dens	50 MWth/ft3
Reactor Size	2 ft3
Reactor Dens	5 X water
Reactor mass	624 lb
Vessel Wall Equiv	3 inches
Repr Vessel L/D	2
Repr. Vessel Diam	1.084 ft
Repr Vessel Len	2.168 ft
Vessel Wall Dens	3 X water
Vessel Mass	345.435 lb
Pump Mass	33.599 lb
Nozzle mass	15 lb
Subtotal	1018.034 lb
Integration mass	356.312 lb
Bare reactor engine	1374.346 lb
Shield thick	6 inches
Shield Dens	3 X water
Shield mass	148.249 lb
ready to Inst'l	1522.594 lb
Inst'l mass	380.649 lb
Installed engine	1903.243 lb

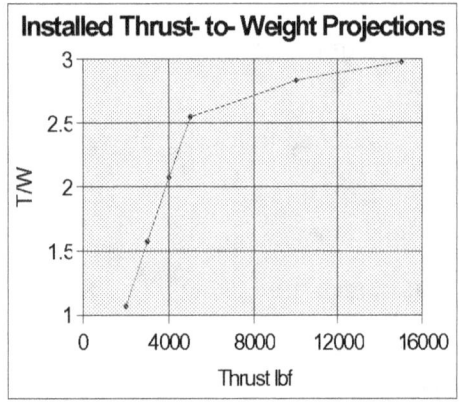

Figure 3-1: Estimated performance for 3 klbf (13.3 kN) NTP

3.3 Specific Observations Regarding Performance Estimates

The NTP inputs received from technologists during IISTP Phase I represent very optimistic technology performance compared to comparable inputs for other propulsion technologies. It is well known to most reviewers that these performance estimates are very optimistic, since there is a long history of NTP technology investigations in the U.S. and, apparently, in Russia.

Conservative estimates would lead to stage masses on the order of 20% more for the small stages appropriate to robotic outer planet missions. Conservative estimates do not have a severe effect on the scores for the NTP option.

Fuel Form and Fuel Temperature; Isp

The fuel form (of the optimistic estimates) is small cylinder elements consisting of cylindrical rolled, perforated tungsten metal foils with uranium oxide fuel. There are a few dozen of these making up the entire reactor. The propellant flow is from the outside of the cylinder into a central axial flow cavity, which discharges into a small nozzle for each fuel element. These nozzles presumably dump into a large diverging nozzle which continues to expand the flow.

The effect of the rolled perforated foils is to create a porous fuel element with very large area per unit volume of contact with the hydrogen propellant. This, it is argued, permits very high power density (and hence low mass) for a given thrust level. The fuel is operated at about 3000K and the delta T between the fuel temperature and the hydrogen propellant temperature is claimed to be about 50K.

Issues associated with this design include:

(1) An inverse trade exists between heat transfer area per unit volume, and sensitivity of the fuel form to hydrogen corrosion. Thin foils as proposed will be subject to serious degradation even at low corrosion rates. I don't believe the corrosion rate for tungsten/UO_2 fuel at these temperatures is well established. There was work done on tungsten-based fuels during the NERVA days, but I don't know how much testing was accomplished, nor at what temperature. There may be more recent Russian data, but I don't know how reliable it is. Russian technology claims are sometimes driven by desire for funding, as is true in the West. The Russians tend to be stingy with raw data.

(2) At high power densities it becomes difficult to match propellant flow distribution with neutron flux distribution which controls power level. Achieving a low and uniform delta T between fuel and propellant requires a very good match everywhere. If propellant flow exceeds proportional neutron flux locally, the hydrogen temperature is reduced. If propellant flow is low, the hydrogen and fuel temperature go up and the fuel temperature limits are exceeded. There are limits to the accuracy with which neutron flux can be predicted.

A more conservative design would reduce power level per unit fuel mass, and reduce design propellant temperature to give more margin between nominal fuel temperature and failure temperatures. A very conservative design would use the NERVA-type graphite fuel form, hexagonal rods with axial hydrogen flow passages (19 per rod for Nerva; this number can be altered), which has been extensively tested. A more conservative fuel temperature would be about 2700K, or if graphite were used, 2500K. Corresponding Isps are about 875 and 850. (The NERVA was estimated at about 800 but it used a hot bleed cycle for turbopump drive. Today we would consider an expander cycle to be state of the art, and this increases Isp at a given fuel temperature, because all of the hydrogen flow is heated to full operating temperature.) A more conservative engine mass would be about twice the Mitee estimates.

3.4 Sensitivities

Figure 3-2 presents a typical graph of achievable vacuum Isp for a nuclear thermal rocket engine as a function of hydrogen temperature. Fuel maximum temperature would typically be at least 100 K greater than the average hydrogen temperature.

- Isp is limited by the temperature of hydrogen flowing through the nozzle.
- The temperature is limited by the maximum material temperature of the reactor core. Allowances must be made for delta T between the reactor core and the hydrogen, and for "hot channel factors" (there will be hot spots in the reactor at higher temperature than the average core temperature).
- Nerva reactors operated at hydrogen temperatures about 2500K
- 2700K is a reasonably conservative assumption for a newly- developed engine.

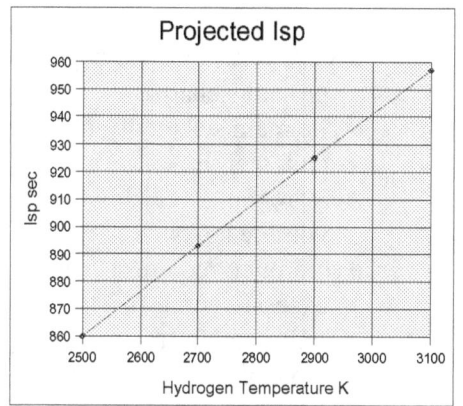

Figure 3-2: Achievable Isp

A nuclear thermal rocket engine is also sensitive to the propellant fraction of its propulsion stage. Like a chemical stage, high performance is achieved when the mass ratio start/cutoff is greater than 2. Table 3-1 presents typical stage mass fraction and limit delta V capability. The figure assumed an aluminum hydrogen tank. As noted, a graphite composite hydrogen tank would decrease inert mass and increase limit delta V capability by about 2 km/s.

4.0 Mission Analyses

Three potential missions were briefly analyzed: Solar System Escape (to 200 a.u.), Mars Sample Return, and Europa Lander. These analyses were only sufficient to indicate general performance potential for NTP.

Solar System Escape ... A solar system escape has been achieved by at least 4 spacecraft: 2 Pioneers and 2 Voyagers. These achieved escape by a Jupiter gravity assist. After 30 years of flight, the Pioneers at about 80 a.u. are approaching the heliopause, the place where the solar vicinity environment transitions to the interstellar environment. The design challenge for new technology in-space transportation is to reach 200 a.u. in 20 years, without a Jupiter assist. Figure 4-2 summarizes the performance requirements. The upper curve shows trip times to distances from 10 a.u. to 200 a.u. as a function of solar C3. The lower curve shows solar C3 as a function of high-thrust delta V, starting from an Earth escape condition. An Earth vicinity gateway could also be used, as these are very close to Earth escape energy. This mission is just barely within reach of an NTP system. The required delta V is nearly too much for a system with Isp less than 1000 seconds. However, it could be done. NEP may be a better choice, but probably represents a more expensive development.

Table 3-1: NTP Stage Sensitivity

- Sensitivities for the engine were displayed on previous charts.
- The main sensitivity for the vehicle is hydrogen tank weight.
- Hydrogen is very low density, requiring large volume tanks.
- Hydrogen requires cryogenic insulation to avoid excessive boiloff.
- Tank fractions (tank mass/liquid hydrogen mass) typically range from 15% to 25%.
- This limits the mass ratio achievable by a nuclear stage, and therefore the delta V deliverable by a single stage
- A sample calculation for a 3000-lb-thrust stage with 15,000 lb propellant load indicated a limit delta V (no payload) about 12.5 km/sec
- If the tank were graphite composite instead of aluminum, about 2 km/sec was added.

Estimate of Vehicle Characteristics	
Propellant Load	15000 lbf
Proellant Dens	4.4 lb/ft3
Propellant Vol	3647.727 ft3
Tank ends b/a	0.7
Fu Tank Ftu	55000 psi
FSult	1.5
Tank diam	14 ft3
Tank Pressure	35 psia
Tank Matl Dens	0.103 lb/in3
Insul Dens	6.00 lb/ft3
Insul Thick	1 in
Insul Area Dens	0.5 lb/ft2
Tank ends e	0.714
Ends surf a	442.984 ft2
Ends Vol	1005.729 ft3
Cyl Vol	2641.999 ft3
Cyl Area	153.938 ft2
Cyl Length	17.163 ft2
Cyl Surface Area	754.857 ft2
Total Area	1197.841 ft2
Tank Wall	0.08 in
Tank Ideal Mass	1424.54 lb
Insul Mass	377.428 lb
Tank Actual Mass	2514.238 lb
Tank Mass Fractio	0.168
H2 press mass	118.994 lb
H2 Resid Mass	150 lb
Prop Delivery Sys	135.814 lb
Total propulsion sy	4822.289 lb
Total mass	19822.289
Propellant Fraction	0.7567
Limit Delta V	12.377 km/s

Mars Sample Return ... The Mars Sample Return case was similar to the case described in the SEP assessment. The return was assumed accomplished by chemical propulsion, and the NTP is used to launch the system from low Earth orbit to trans-Mars injection. This places a significant delta V on the NTP stage and results in a much lower mass mission than the all chemical system, but more massive than the SEP case.

Curves to the right show trip times and delta Vs versus solar C3, starting from an Earth escape condition.

A single 3K NTP engine with staged cluster tanks can achieve between 20 and 25 km/sec (3 tanks and 7 tanks respectively). Burn times are 3.7 and 8.7 hours.

Distances 100 to 200 AU are achievable in 12 to 30 years.

Such a vehicle could be assembled at Earth-Moon L2 from a few EELV-H launches.

Cost of this implementation can be traded versus something like a high-power NEP where the propulsion system will be more expensive, but the launches (presumably fewer required) less expensive.

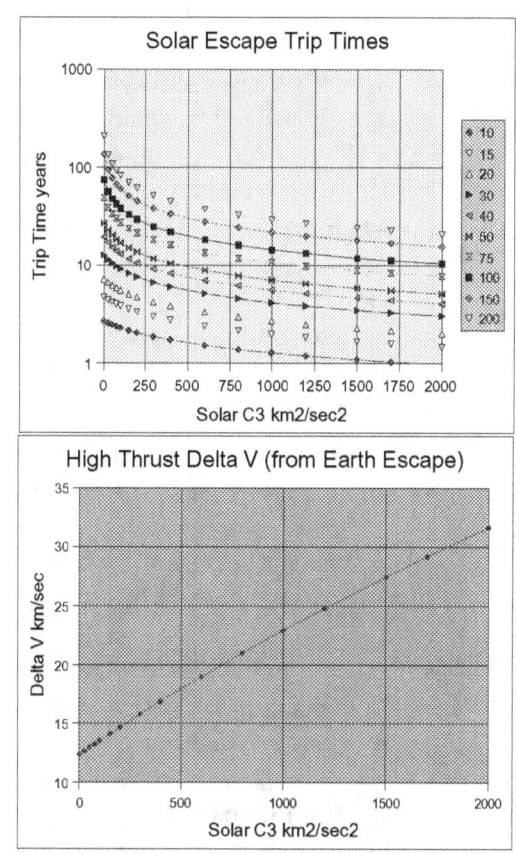

Figure 4-1: Solar System Escape Mission Analysis

The Mars sample return mission is summarized as follows:

- Mission profile assumed:
- Launch from LEO by NTP to Mars transfer C3 15 km2/sec2
- Lander does direct entry as before
- Return vehicle does aerocapture at Mars.
- Assume aerobrake plus maneuver propulsion mass penalty 25%.
- Ascent vehicle rendezvous with return vehicle in Mars orbit; payload carrier transferred; return vehicle performs TEI and capture in LEO upon return to Earth.
- NTP does TMI of entire vehicle from LEO to Mars transfer. NTP stage disposed by Mars gravity assist at arrival.
- Assume return vehicle 6320 kg as for chemical vehicle.
- Return vehicle ready to capture at Mars 7900 kg.
- Entire package ready to launch to TMI 7900 + 2500 = 10,400 kg
- Assume NTP g loss 200 m/s
- NTP delta V = sqrt(2*μ/r + C3) + 0.2 - Vc, = 11.64 -7.61 = 4.03 km/s
- NTP mass ratio at Isp 850 = 1.62
- Required NTP propellant load 17,500 lb.
- Total launch mass to LEO 46,000 lb, in range of a single EELV-heavy

For the Europa Lander mission, operations at Jupiter can use gravity assist by Jupiter's large moons. The mission is summarized as follows:

- Capture into 100,000 x Ganymede elliptical orbit ~ 2 km/s
- Note that a moon near periapsis can be used to pump apoapsis and a moon near apoapsis can be used to pump periapsis.
- Use Ganymede to pump periapsis up to Europa (also increases apoapsis somewhat) and use Europa to keep apoapsis pumped down to near Ganymede.
- Ideally achievable Vinf at Europa is 1.69 km/s; actual probably closer to 2 km/s. Then capture and landing delta V ~ 3 km/s.
- Using Isp 325 sec and propulsion propellant fraction yields 2050 kg spacecraft for 250 kg landed bus mass.
- 3K nuclear stage from Earth escape to C3 ~ 90 delta V = 9.5 km/s (assumes nuclear start at high altitude); mass to escape 30,000 lb
- 3K nuclear stage from LEO to C3 ~ 90 delta V = 7 km/s; mass to LEO = 19,000 lb

IISTP Phase I Final Report
September 14, 2001

5.0 Design Strategies and Approaches

- Small engines to simplify ground testing
  - 3K thrust for robotic missions
  - 15K thrust for human missions with clustering as needed
- Core temperatures compatible with required burn times and number of restarts
- No reason to deviate from traditional NTR design approach
  - Nerva-type fuel element design, perhaps more holes
  - Feasibility questions (sintering, etc) with particle-bed design
  - Don't need higher power density for small reactors
  - Evaluate Russian twisted-ribbon designs
  - Reflected cylindrical design; reflector control drums
  - Hydrogen cooling of vessel, structural parts and controls
  - Shadow shield
  - Adequate materials testing (in hot hydrogen) for new reactor materials; or use Nerva materials within their temperature limits
- Probably should use expander cycle rather than bleed cycle for pump power.
- Redundant pumps for human missions to minimize risk of loss-of-coolant (LOC) accident
- Hydrogen CFM for NTP is the same as for cryo chemical

IISTP Phase I Final Report
September 14, 2001

6.0 Desirable Engine Characteristics

- Highest reactor operating temperature consistent with desired life and reasonable design margins.
- Lifetime 1 to 10 hours depending on mission.
- Reactivity control range does not extend into prompt critical region.
- Large area ratio nozzle for higher Isp.
- Flat neutron flux profile probably requires fuel loading gradations in small engines.
- Rigorous quality control needed because flight engines will not be acceptance tested. Pumps can be run on a pump stand.
- Engine fitted with neutron poison (e.g. wires in coolant passages) to preclude criticality in event of water immersion on launch accident.
- Usual design includes a shadow shield for protection of payload. Hydrogen tank provides additional protection, especially when filled with hydrogen.

7.0 Test Facility and Development Requirements

Development of liquid rocket engines, regardless of the heat source, has historically required a lot of testing. The F-1 engine went through about 50 builds and 1000 firings to reach flight qualification. This was, of course, a hardware-rich program. The SSME, which was hardware-lean, went through about a dozen builds and hundreds of firings. Since a nuclear rocket does not involve combustion devices, it may require fewer. We have no valid history, since no nuclear rocket ever went to qualified status.

Of the options offered for engine development, the idea of doing development in a few flight tests is completely unrealistic. The other options are to test using existing man-made downholes at a former underground weapons test facility, and building a new complete containment facility at INEL. A new facility is probably required, but the downhole approach may work and should be evaluated. Either case requires a containment structure to deal with loss-of-coolant or other engine failures; this is apparently proposed for both options.

My guess is that a nuclear rocket program would go through something like the following, assuming enough design conservatism to avoid major development problems:

(1) Critical assembly ... neutron flux, criticality measurements, and control drum effectiveness at low power, no propellant flow. No special facilities required.

(2) Turbopump development ... The turbopump would be developed by non-nuclear testing, with simulated heat source for delivering hot hydrogen turbine drive and simulated thrust chamber for discharge flow resistance. There are several existing test facilities capable of this. Some special test equipment may be required, but not a schedule or cost issue.

(3) First build: Neutron flux and flow distributions at power, operating with facility-fed hydrogen (no turbopump). This requires a nuclear test facility. Tests would begin with start transients, continue at reduced power and temperature, and increase in power, temperature and duration as data are collected. Frequent hardware inspections would be required to assess local hot spots and any other damage or deterioration. Most of these could be implemented with robotics by snaking a fiber-optic viewer up through the nozzle exit. Some may require engine removal and tear-down in a robotic hot cell. These capabilities are well within the state of the art of robotics and pose no special issue except cost of equipment development.

(4) Second build: Correcting deficiencies in the first build and continuing facility-fed hydrogen tests, attaining design duration and power level. Testing would continue through design core life, assuming no fuel failures to that point. Facilitization and

procedures same as first build. Following test series completion, the core would be completely dismantled and inspected for damage or deterioration.

(5) Third build: integrated engine tests ... the turbopump is added, and integrated engine testing begins. Testing proceeds through start transients to progressively longer burns. Engine inspections are conducted as for the first and second builds. Turbopump performance is closely monitored, but unless anomalies or excursions beyond design operating conditions occur, the pump is not removed or inspected until completion of the test series. Following completion, the engine is completely dismantled and inspected for damage or deterioration.

(6) Fourth build: Integrated engine tests continue to explore the operating envelope specified by the engine specification. Performance closely monitored, but only in-place inspections required unless anomalies are encountered. Following test series completion, the entire engine would be completely dismantled and inspected for damage or deterioration.

(7) Fifth build: Qual test engine. The qualification test series would be performed. Performance closely monitored, but only in-place inspections required unless anomalies are encountered. Following test series completion, the entire engine would be completely dismantled and inspected for damage or deterioration.

(8) Sixth build: First flight engine. No acceptance test is performed on the integrated engine. The turbopump would undergo an acceptance test before installation on the engine.

The test facility includes exhaust containment, which could be in a new facility as in the INEL test concept, or in the down-hole facilities as proposed for Nevada testing. The down-hole concept needs to be examined to ensure it is capable of satisfying a test program as described above. My guess is that the test program currently conceived for this facility involves much less testing than described above. Figure 7-1 presents some rough calculations on a containment facility concept

	Thrust Levels	
	3000 lbf	15,000 lbf
Hydrogen throughput (kg), 1 hr at 890 Isp	5504	27520
Volume (m^3) @ 1 atm pressure & 1000 K[1]	224,100	1.12x10^6
Length of 5- m dia duct (km)	11.4	57
Volume if water spray to 400K (water vol negl)	89,600	448,200
Length of 5- m dia duct (km)	4.6	23
Water flow rate required, gpm	250	1250
Volume if LOX injected to burn hydrogen to water and then all is condensed by water spray[2]	8360	41,800
LOX flow rate required	12.1	60.7
Water flow rate required, gpm	17,102 (600 hp)	85,513
Length of 5- m dia duct (km)	0.5	2.5

1: Assumed it will cool to about this temp due to heat transfer to duct wall
2: Assumed 5% H$_2$ gas residual, plus water volume

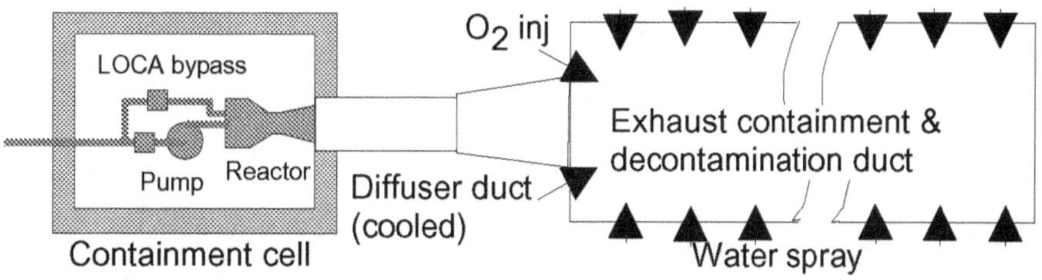

Figure 7-1: Exhaust Hydrogen Containment Concept

In addition to exhaust containment and a containment structure to deal with reactor or turbopump (loss of coolant) failures, the test facility must provide thorough instrumentation, robotics for engine in-place inspection, and engine removal and re-installation, and one or more robotics hot-cells for engine disassembly, inspection, and re-assembly.

8.0 Cost Considerations

- Non-recurring:
    - Main cost of nuclear thermal propulsion is re-creation and maintenance of the R&D institution required.
    - Test facilities are significant cost.
    - Potential cost of delays due to dealing with environmental impact issues and political opposition to nuclear propulsion in space.
    - These are not amenable to CER analysis because there is no applicable history.
- Recurring:
    - Engines expected to be $100 - $150 million each
    - Costing should be possible with CERs
    - Balance of vehicle is conventional cryo vehicle
    - May want to add a little to avionics for rad hardening.
- Operations:
- Similar to other high thrust systems
- Perhaps added systems safety costs
- Post-mission reactor disposal but this appears to be a minor cost.

9.0 Strategies for Technology Advancement and Mission Readiness

- Certify fuel form and materials by thorough testing
- Small engine first to minimize costs of problems
    - Intent to use NTP on HEDS missions is sufficient reason to do small engine first.
- May want first in-flight use to begin at Earth escape to minimize safety issues
- Thorough ground test program in effluent-containment & decontamination facility
- See no reason for a technology flight test.
- NTP engine is similar to chemical rocket engine
- High confidence it will work in space if it works in ground test facility
- Flight engines cannot be acceptance tested because they become radioactive. Can and should do pump "green runs".
- May be a useful functional end-to-end test with hydrogen flow but no nuclear power. Check all valves and controls, pump spin-up.

10.0 Conclusions

- Applicable to most mission categories
  - Inner solar system complex profile
  - Outer solar system simple profile
  - Outer solar system complex profile
  - Beyond solar system
  - HEDS Mars and asteroids

- Technology well-understood in principle
  - Rover test program
  - Mature analytical capabilities
  - Significant public safety and environmental issues
- Effluent containment and de-contamination test facilities are feasible
  - Containment of loss-of-coolant accident required; may be main cost impact on test facility
- Launch "virgin" reactors; not significantly radioactive
- If we want NTP in the stable, do small engine first
- Non-recurring costs require careful evaluation; not a CER problem

IISTP Phase I Final Report
September 14, 2001

SEP Independent Assessment
Gordon Woodcock
Gray Research, Inc.

1.0 Introduction and Purpose

This assessment was conducted as part of Gray Research support to the MSFC In-Space Integrated Space Transportation Planning activity. The purpose of the assessment was to develop an independent understanding of the performance potential for solar electric propulsion, and the technology characteristics that would best serve utilization of solar electric propulsion for future space missions.

2.0 Solar Electric Propulsion Mission Considerations

Basic Principles - Basic principles for electric propulsion were described briefly in the accompanying NEP independent assessment and are not repeated here. Mission ideal delta Vs for nuclear and solar electric propulsion are similar, and the range of optimum Isp is also similar. At low power, up to 100 kWe, solar power systems are predicted to exhibit less mass per unit power than nuclear systems, and nuclear systems have the advantage at high power levels.

A major difference between the systems is that nuclear electric power systems are not dependent on the Sun as an energy source and therefore can deliver power and propulsion in the outer solar system, where solar electric propulsion is ineffective to useless.

Mission Applications - Principal mission application considerations are:
- Electric propulsion is not useful for landing on or ascent from object with significant gravity.
- Chief performance limitation is mass required to convert power and produce thrust
- An SEP must point solar arrays to Sun
- SEP must be at a reasonable distance to Sun for thrust power. Array output may go to zero at low light levels. Thus, SEP may not even produce housekeeping power in outer Solar System
- Electric storage is not a reasonable thrust power option for shadowed periods. It's OK for bus housekeeping. Should assume propulsion off during shadowed periods.

Table 2-1 presents a summary of mission types reviewed by IISTP, and expected constraints and applications for solar electric propulsion.

Table 2-1: Potential Applications for Solar Electric Propulsion

Mission Type	Expected Application/Utility
LEO station keeping	State of the art. Possible problems with high precision due to shadow periods
LEO to GEO or libration point	Spiral through van Allen belts produces significant radiation dose
GEO or libration point station keeping	State of the art. May not be suitable for close proximity formation flying due to high-velocity jets
Pluto Flyby	Outer solar system requires threshold power-to-weight (described later)
Europa or Neptune Orbiter	Not suited for maneuvers at Jupiter or Neptune but chemical + aeroassist + gravity assist suffice
Mars Sample Return	May require long life (up to 4 year mission); mission described on later charts
Kuiper Object Rendezvous	Not suitable; can't produce thrust at destination for rendezvous
Beyond solar system	Can reasonably be expected to achieve solar system escape, but transit to ~ 200 AU in reasonable time probably not achievable.
HEDS lunar	Useful for cargo; not for crew due to long trip time
HEDS Mars/Asteroid	Need high power 5 - 20 MWe. There are significant technology challenges to achieving multi-megawatt solar electric systems, although probably less so than for high power nuclear electric systems. Weak Stability Boundary gateway basing may be appropriate.

3.0 Issues

3.1 Integration Factors

Area: Useful power levels range from 1 to 2 kWe for station keeping, to 10 to 50 kWe for robotic planetary missions, to 5 to 20 MWe for human planetary missions. These power levels are referenced to Earth's distance from the Sun. High-performance solar arrays produce 300 to 400 watts per square meter. The corresponding areas are 3 to 6 sq m for station keeping, 30 to 150 sq m for robotic missions, and 15,000 to 60,000 sq m for human missions. These large areas impose packaging and deployment or assembly problems.

Voltage and Power Distribution: Up to about 25 kWe, conventional solar array voltages (28 to 160 V.) are practical. The International Space Station uses 160 VDC distribution for up to about 100 kWe. At 160V, 100 kWe represents 600 amps. Conductor losses can become severe, or conductors massive, as power, current and distribution distance increase. At some point it becomes imperative to employ advanced high-voltage power distribution methods.

Flight Control: Large area structures introduce control issues. Ordinarily, a SEP stage does not need to be very maneuverable, but the thrust vector must be accurately directed. Large area structures are likely to have low natural frequencies and a large number of significant flexible modes. Attitude control stability must be maintained in the presence of these flexible modes. Electric thrusters need gimbal capability to maintain thrust through the vehicle center of gravity and provide roll control. Given the low thrust, the gimbal motion can be slow.

3.2 Ranges of Achievable Mass/Power Performance

Power in the jet is $fu/2$ where f is thrust in newtons and u is jet velocity in m/s. We need f/m to be at least a few x 10^{-4} for some missions 10^{-3}. (Note 10^{-3} is 30 km/sec per year.) Then $p/m = f/m\ u/2$, and for $f/m = 10^{-3}$, p/m is about Isp/200 in watts/kg. If efficiency of conversion of electric power to thrust is 60%, then p/m electric needs to be about Isp/120. Isp/200 to Isp/100 is a representative range, for the spacecraft as a whole, with payload and propellant.

An optimized electric propulsion system tends to be about 1/3 propulsion system, 1/3 bus and payload, and 1/3 propellant. For the power system, one could then quote the power-to-mass ratio as Isp/60 to Isp/30. This mass split gives a mass ratio 1.5. Therefore, the rocket equation would specify that Isp should be about 250 times the mission ideal delta V in km/sec.

Example: Mission ideal delta V 12 km/s, Isp = 3000 sec. Since we assumed power system = payload mass, by this rule of thumb, at 3000 Isp we want 50 to 100 watts per kg. (10 to 20 kg/kWe) If the payload is 100 kg, the power is 5 to 10 kWe.

For most missions, the velocity needs to be delivered in less than 2 years as a maximum. Multiply 10 kg/kWe by 730/132 to obtain 55 kg/kWe as a rough maximum acceptable mass/ power ratio. For human missions to Mars and return, on opposition-like profiles (i.e. fast round trips) the calculated power duration of 132 days is already about as long as we would wish to entertain, so for these missions, the mass/power ratio should be less than 10 kg/kWe. Only very low power/mass solar electric systems are likely to achieve less than 10 kg/kWe. Examples described in this assessment are not that high in performance.

3.3 Selection of Isp

An optimum Isp occurs for electric propulsion systems. If the mission delta V is not dependent on Isp (for example, delta V for a low-thrust spiral from LEO to GEO is only weakly dependent on Isp), a simple optimization may be performed as indicated in Figure 3-1. For planetary missions the delta V is usually strongly dependent on Isp and this simple optimization is inaccurate. The principle still holds.

- An optimum Isp occurs because if Isp is too high, powerplant mass dominates; if too low, propellant mass dominates.
- If mission ideal delta V is not a function of trip time, a closed- form equation exists:

$$M_L/M_0 = 1/\mu - x[(\mu - 1)/\mu]u^2/(2T)$$

where M_L/M_0 is payload fraction
μ is rocket equation mass ratio
x is mass/power in kg/watt of jet power
u is jet velocity in m/s
T is burn time in seconds

If mission ideal delta V is a function of trip time (the usual case) the optimum Isp must be found by trajectory analysis, but the general principle still applies.

Example ...
12 km/s delta V in 150 days
30 kg/kWj (jet power; about 18 kg/kWe)

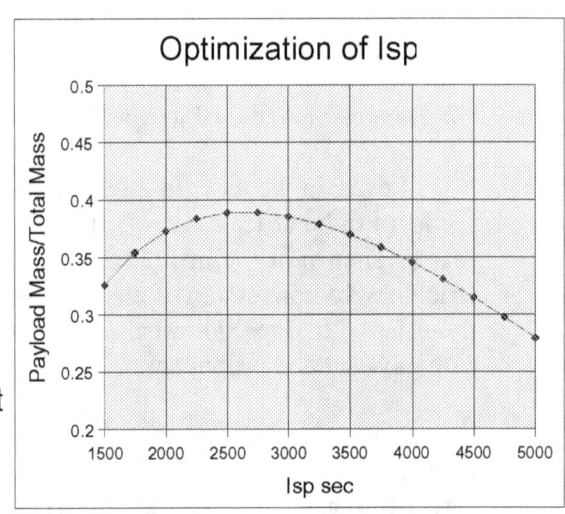

Figure 3-1: Optimization of Isp – Simple Example

3.4 Sensitivities

For the SEP assessment, we considered sensitivity of thruster and power processing efficiencies. More efficient systems can afford to be more massive while still delivering equal performance because the higher efficiency delivers more propulsion from the solar array mass. A representative high-performance system would exhibit mass/power about 15 kg/kWe, For this system the thruster and power processing efficiency/mass trade is illustrated in Figure 3-2.

Sensitivity curves are based on an advanced solar electric propulsion overall electric power/mass ratio 15 kg/kWe at 65 kWe. Advanced gap solar cells are assumed. Break-evens maintain constant

Item	Nominal Mass/Power	Nominal Efficiency
Blanket	3.6	(25)
Blanket	2	N/A
Main power	1	98
Power proc. &	4	95
Thrusters	4	65
Propellant	0.4	N/A

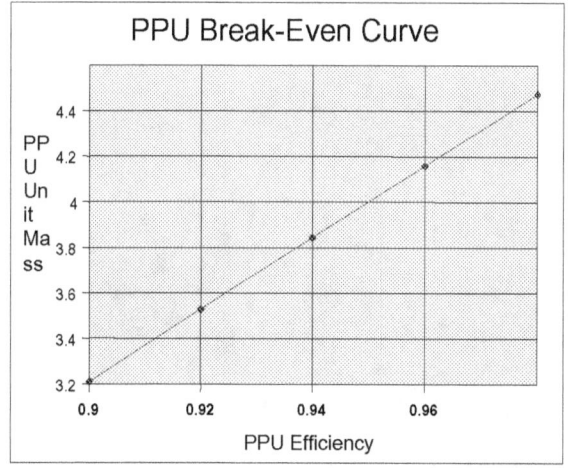 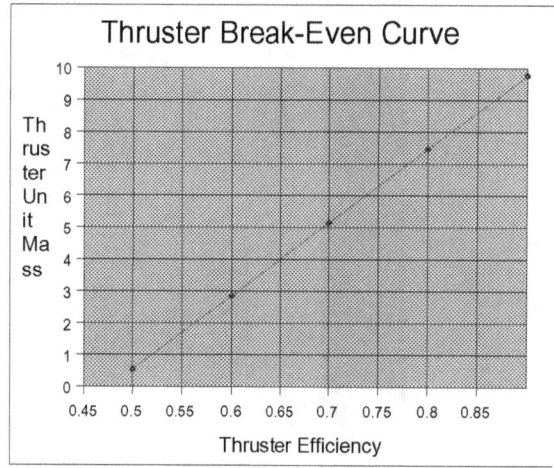

Figure 3-2: Performance Sensitivities

IISTP Phase I Final Report
September 14, 2001

4.0 Mission Analyses

Two mission examples were considered: One was solar system escape, and time to Pluto distance (here taken as 40 AU although Pluto is presently nearer the Sun at about 30 AU). Solar electric propulsion, because it is dependent on the Sun for power, exhibits reasonably effective trajectory performance for solar system escape, i.e. less than one revolution about the Sun and attainment of escape energy in about a year, above a certain power-to-mass ratio, and poor performance. i.e. more than one revolution and many years to attain escape energy, below that value. The transition is relatively rapid near the critical power-to-mass ratio as illustrated in Figure 4-1. The charts in the Figure are the result of numerically integrated but not optimized trajectories. Illustrated performance can be improved by use of Venus or Earth gravity assist.

The second case considered was Mars Sample Return, which is a round-trip mission. In the case evaluated, the SEP was assumed launched by an ELV to positive C3 re Earth. Its profile is (1) transit to Mars carrying the lander, (2) capture and spiral down to a low Mars orbit where it picks up the Mars sample from the Mars ascent vehicle, (3) transit to Earth, and either (a) releases the sample vehicle for a direct Earth entry, or (b) spirals down to a low Earth orbit for Shuttle pickup.

SEP requires a threshold thrust-to-mass ratio to achieve solar system escape, about 10^{-4} g. Below this value, power drops off too fast with distance from the sun. The curve to the right shows time to reach C3=0 (re Sun) beginning at Earth distance. T/M was set at 0.001 N/kg (approximate T/W 10^{-4}g) at Isp 3000. At constant power, as Isp increases the T/W decreases. A 12% increase in Isp doubles the time to reach escape. Solar system escape capability is a reasonable test for outer planet mission feasibility.

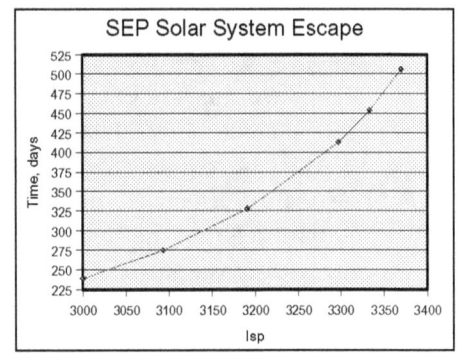

A representative SEP system at 65 kWe would have the following characteristics:
⋇ Efficiency 60%; jet power 39 kW
⋇ Isp 3000 sec; thrust 2.6 N
⋇ Mass 2600 kg @ Earth escape; Delta-III class
⋇ Propellant load 1040 kg (to solar system escape).
⋇ Spacecraft and payload 1560 kg
If the electric propulsion system has specific mass 15 kg/kWe, its mass is 1000 kg, leaving 560 kg for other spacecraft bus functions & payload. 15 kg/kWe is in the achievable range, e.g. with 200 W/kg solar array.

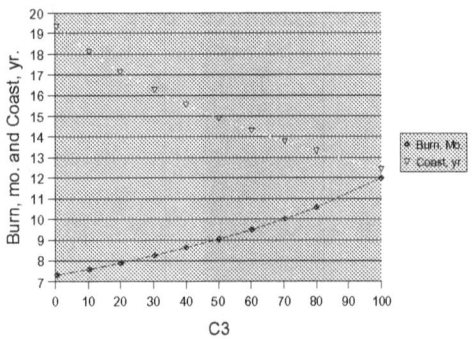

For about the same performance, a solar sail would be 600 m square at 2 g/m² (very light, beyond near-term state of the art). It could use a smaller launcher, Delta II instead of Delta III.

Figure 4-1: Solar System Escape Mission Performance

A Mars sample return mission description was contrived as summarized below for purposes of SEP performance analysis and comparison to chemical propulsion. No representation is made that this Mars sample return example is a recommended mission.

Payload carrier bus
- Carries Mars samples, includes ample protection provisions
- Performs rendezvous and docking with SEP interplanetary bus after ascent from Mars to low circular orbit
- Assumed 100 kg for 2 to 5 kg sample payload

Mars ascent
- 2-stage solid or Mars surface storable propellants.
- Solid Isp assumed 290, propellant fraction 0.9
- Equivalent liquid Isp 325, propellant fraction 0.836
- Did not assume Mars surface propellant production, but should be evaluated.
- Ascent vehicle gross mass ~ 900 kg including payload carrier bus.

- Lander

- Assumed to carry 200-kg rover (or 2 - 100 kg rovers) plus ascent vehicle
- Assumed to use entry aeroshell plus parachute plus terminal descent propulsion.
- Assumed aeroshell 12%, landing propulsion 11%, parachute 15%, and structure & bus 28%, respectively, for lander inerts as fraction of total landing weight.
- Total landed payload 1100 kg; estimated entry weight 2500 kg.

The basis for the SEP analysis was delivery to Mars, 2500 kg, return to low Earth from Mars, 100 kg.

Figure 4-2 presents a summary of the estimated mission performance. Delta Vs were not obtained from specific trajectory analyses but from generic characteristics of these trajectory types. The initial mass is within the capabilities of existing ELVs. The profile feature of a spiral down to low Earth orbit is attractive from the point of view of planetary back-contamination.

We assume 1000- kg SEP as before, with Isp 3000 and jet power 39 kW.
Thrust 2.65 N mass flow 9×10^{-5} kg/s
We assume launch to C3=10 - 15 km^2/sec^2
Propellant tanks are assumed 10% of propellant capacity. One tank is used for Earth spiral- in and a second tank for all prior maneuvers.
We assume the SEP delivers half power (half propellant flow) in Mars vicinity.

Maneuver	Delta V km/s	Mass ratio	Propellant required	Burn time (days)	End mass	Start mass
Mars arrive	5	1.185	1008	220	5442	6450
Mars spiral- in	3.5	1.126	330	85	2612	2942
Mars spiral- out	3.5	1.126	304	78	2408	2712
Mars- Earth	10	1.405	694	116	1714	2408
Earth spiral- in	7	1.269	304	39	1130	1434

The lander separates at the end of the Mars arrive maneuver. The propellant tank for the last maneuver weighs 30 kg; the propellant tank for the remainder of maneuvers weighs 234 kg. Total burn time is 538 days (12,912 hours). In view of the long burn for Mars arrive, the Earth-Mars trajectory probably needs to be type 2.

Figure 4-2: Mars Sample Return SEP Performance Estimate

A comparable chemical mission performance estimate is as follows:

- Launch to Mars transfer as for SEP.
- Mars Arrival: assume C3 = 15; capture into 500 km circular orbit.
- Mars Departure: C3 = 15; Earth arrival C3 = 20.

- Earth arrival: Consider direct entry (not comparable) and capture into low Earth orbit. For direct entry assume return payload 200 kg.
- Mars lander is same as for SEP. Note that chemical mission must dispatch the lander at Mars approach C3=15 rather than C3~0 as for the SEP case.
- Assume ascent stage same so we don't have to rerun ascent analysis.
- Assume Isp 370 & propellant fraction 0.82 for space propulsion stages (rough estimate for LOX-methane).
- Return mission: DV TEI = 2.77 km/s mass ratio 2.14 W = 2.87 Bus mass 200 kg
- Return vehicle loaded mass = 2.87*(200 + 200) = 1148 kg
- Mars capture delta V = 2.77 km/s
- Captured mass = Return vehicle = 1148 kg
- Begin capture mass = 3295 kg
- Lander delivered = 2500 kg as before; total launch mass = 5795 kg to C3 = 15.
- If capture to LEO for quarantine (as assumed for SEP), DV = 4.04 km/s; mass ratio 3.04; W = 5.5. Return vehicle is then 6320 kg and total Earth launch is 20,637 kg.

5.0 Technology

Thrusters: Thruster characteristics to enhance system performance are as follows:

- Efficiency ...   Directly affects jet power-to-mass ratio; target > 60%

- Long life ...   SEP missions may require a year or more of operation

- Reliability ...   For mission success, need to reliably deliver long life
- and performance

- Benign failure ...   Should not take down the rest of the system on failure; defeats redundancy of thruster clusters.

- Low mass ...   Target is on the order of 2 kg/kWe

- Simplicity ...   Not requiring elaborate power processing/conditioning

- Cost   ...   Should not exceed "pricey" space hardware ~ $10,000/lb

- Propellant ...   Should be available at reasonable cost; permit efficient
- tankage/storage; not require exotic delivery systems.

There are numerous thruster technologies. While this independent assessment did not spend much time on thrusters, brief observations are in order.

Resistojets and *arcjets* have been used commercially. Both use thermal heating of propellant and thermodynamic expansion through a nozzle. Isp available from these types is too low for planetary mission applications (Isp up to 600 seconds in the commercial versions, which use hydrazine; arcjets could probably reach about 1500 seconds with hydrogen).

Hall thrusters use the Hall effect with a current flowing in an ionized plasma to create an electric field which accelerates ions. Commercial versions have achieved Isps about 2000 seconds at power levels of a few kWe. This Isp is about ideal for operations in Earth orbit and Earth vicinity, but too low for planetary missions. Hall thrusters exhibit efficiencies in the low to mid 50s. Future developments may improve this. Current Hall thruster technology is limited in power to 10 kWe or so. Discussions during IISTP Phase I indicated that Hall thruster power up to about 100 kWe may be feasible, but that attaining significantly higher Isp is not expected.

Ion thrusters Use electron optics (a series of grids, usually 2 or 3) to accelerate positive ions created in an ionization chamber. Electrons are emitted by a separate electron gun, usually called a neutralizer, so that the net beam is charge neutral. Current ion thrusters, commercial and government-sponsored, have Isp in the range 3000 to 4000 seconds at power levels slightly less than 5 kWe. The next logical development step is a 10 kWe ion thruster with maximum Isp somewhat greater. Ion thrusters exhibit efficiency exceeding 60%. Building ion thrusters with power capacity greater than a few tens of kW is expected to prove very difficult.

Several other thruster types are in an earlier stage of technology development. Experimental *magnetoplasmadynamic (MPD)* thrusters have used noble gases and lithium as propellant. Power levels are multi-hundred-kW and up. A wide range of Isp, 2000 to 10,000 seconds, has been discussed. Experimental efficiencies are lower than ion or Hall thrusters, with the lithium devices presently doing somewhat better than the gas thrusters. Continuing research is directed, in part, to increasing efficiency. A *pulsed inductive thruster (PIT)* was developed and tested several years ago by TRW. It uses a high-intensity electrical pulse through a spiral-wound magnet to ionize and accelerate a pulse of gas propellant. Efficiencies in the 50% range were measured. At high pulse rates this device would exhibit high power. The *variable specific impulse magnetic rocket (VaSIMR)* ionizes hydrogen, traps it and heats it (by microwaves) in a magnetic bottle, and through controlled leakage, permits the hot hydrogen to expand through a magnetic nozzle. This device is under development at JSC. It can produce variable specific impulse because the hydrogen heater is independent of the ionizer. In view of the complexity of the device and multiple opportunities for losses (ionizer, heater, leakage of neutrals and ions, and nozzle losses), in this reviewer's opinion, it is likely to suffer from low efficiency. If this thruster works, it will be capable of high power.

Ion thrusters are presently favored for robotic planetary mission electric propulsion (either solar or nuclear) because of their maturity, Isp range, and efficiency. Because of their power limits, large numbers would have to be clustered for use on multi-megawatt spacecraft. If one or more of the high-power-capable thrusters reaches a greater level of maturity and efficiency, it will probably be preferred for multi-megawatt applications.

<u>Solar Arrays:</u> Current solar arrays are mostly silicon at about 14% efficiency, some gallium arsenide at about 18%. New multiple-band-gap technology is now becoming commercially available, at about 28% efficiency (25C AM0). A comparison is shown in Figure 5-1. The performance in watts/kg is for a bare blanket; a complete array (with additional structure, mechanisms and wiring) will probably be, for example, about 220 W/kg where the bare blanket is about 280.

There are a number of alternative array technologies that may exceed this performance. Trough concentrators are now used on commercial communications satellites. Other concentrator configurations have been tested. Various thin-film solar cell types could provide high performance if the substrates and coverglasses are light enough.

Power Processing and Distribution: Solar cells generate from one to about 2.5 volts per cell. Series strings are used to raise the voltage to 28 to 160 volts. While higher voltages could be generated, conduction paths and charging effects in low Earth orbit limit the utility of higher voltages. If the system need never operate in a plasma environment such as low Earth orbit, higher voltages could be used. However, thrusters may need voltages up to thousands, and also need control of the power supplied to them. Array output voltage is affected by operating temperature and other effects such that the fluctuation in array output voltage may be as much as 2:1. Therefore, power processing is generally necessary.

Solar Electric Power State of the Art

Array power/mass performance can be about doubled by slight decreases in materials thicknesses and major increase in solar cell efficiency. Currently available Spectrolab cells are quoted at 26% efficiency and 0.8 kg/m^2
Can use 160- V arrays as for space station.
Power processors can get below estimated 5 kg/kWe at 10 kWe and above, by going to aircraft standard frequencies (440 Hz) or above. That's 50 kg for a 10 kWe processor.

Space Station Array Technology

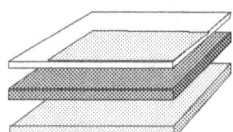

Item	Thick, mil	Density	Mass kg/m^2
Cover glass	5	3	0.375
Cell	8	2.7	0.54
Substrate	10	1.6	0.444*
Adhesives, etc.	1	1.6	0.04
Conductors	0.1	8	0.02
			1.419

At assumed efficiency 13%, power/mass = 123 w/kg

* Array area efficiency 90%

Advanced Array Technology

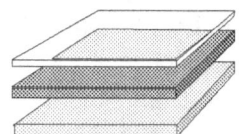

Item	Thick, mil	Density	Mass kg/m^2
Cover glass	4	3	0.3
Cell	4	5	0.5
Substrate	8	1.6	0.34*
Adhesives, etc.	1	1.6	0.04
Conductors	0.1	8	0.02
			1.20

At target efficiency 25%, power/mass = 279 w/kg

* Array area efficiency 95%

Figure 5-1: Illustration of Solar Array Performance

Power processors usually convert dc electricity to ac, use transformers to increase or decrease ac voltage, and then rectify the output at the desired voltage. Alternatively, switching and capacitor ladders can be used to increase dc voltage. Power processing may be a significant part of propulsion system mass. An important factor is how complex the power processing task is to provide the power and power control needed by the thrusters. Low-power systems (few kW) may exceed 10 kg/kWe. Projections at higher power are in the range 1 to 2 kg/kWe.

Design Strategies: Design strategies for achieving high power-to-mass ratio are presented in Figure 5-2. Values for thrusters and power processing in the figure may be pessimistic, depending on technology choice and power level. Achieving high power-to-mass performance is very challenging since mundane factors such as structures, array deployment, launch loads support, and thruster gimbaling mechanisms can add up to severely penalize overall system performance. Achieving high performance from array, power processing and thruster technologies is only half the battle, and the battle is likely to be lost on the other half.

Design strategies for high power may be summarized as follows:
- Multiple array wings and/or panels, like Space Station
- Distributed power processing with conversion to high-voltage 3-phase AC at many points on the array and conversion to thruster power at the thruster location
  - For example, each processor might be 50 kWe, i.e. 20 processors per megawatt.
- Multiple redundant parallel systems, for example a 10-megawatt SEP might be an assembly of 10, 1-megawatt systems.
- High power thrusters, 100 kWe to 1 MWe (but efficiency is still important!)
- Assembly in space, at Space Station or other shuttle-compatible orbit, or robotic assembly at a high-energy location such as a libration point gateway.
- Boost to ~600 km by a low-thrust chemical tug to get out of high drag area
- High-power SEP (such as for Mars transportation) based at high-energy location such as EML2, serviced from LEO by chemical propulsion vehicles.

- High-efficiency multiple-band-cells
- Higher array voltage as for Space Station
- High performance power
- High efficiency thrusters
- Lightweight structures and components
- Find a way to design solar array support and deployment structure that it does not carry launch loads (for example, provide an auxiliary structure as part of the SEP and payload

Representative power to-mass breakout

Item	kg/kWe
Blanket	3.6
Blanket support/deploy	2
Main power conductors	1
PPU/control	4
Thrusters	4
Propellant Delivery System	0.4
Total	15.0

Note: Propellant tanks separately accounted

Figure 5-2: Design Strategies for Power-to-Mass

6.0 Cost Considerations

The SEP unit cost needs to be no more than savings for eliminating, or reducing size of, a launch vehicle. Therefore, our hypothetical 65 kWe SEP should be in the $100 million range.

Solar array @ $1000/watt	$65 million
PPUs @ $5000/kg & 4 kg/kWe	$1.3 million
Thrusters @ $20,000/kg & 4 kg/kWe	$5.2 million
Total	$71.5 million

(Targets representative of commercial space hardware.)

- Leaves about $30 million for integration cost.
- Development cost is typically 5 x first unit cost; suggests a target < $400 million.
- All, of course, after technology has been advanced to Technology Readiness Level (TRL) 6.

7.0 Strategies for Technology Advancement and Mission Readiness

- Solar cell technology appears to "be there". Minor improvements, lightweighting, deployment for larger systems, might be sought.
- Do a systems design study to confirm achievability of useful performance and cost targets, and obtain sensitivity data similar to slide 6. If OK, continue.
- Select ~ 3 thruster technologies based on estimated power-to-mass and efficiency. Others continue research status
- Develop prototype power processors for each thruster type.
- Bring thrusters and processors to TRL 6
- Build and fly a 10-kWe (or thereabouts) flight testbed (this would be next-generation technology beyond Deep Space 1).
- Build a 25 to 65-kWe (or thereabouts) lead customer SEP vehicle and fly it on a real mission.
- Instigate ground-test research into 3ϕ AC 400 Hz (or selected frequency) distributed DC-AC converters and power collection networks for multi-megawatt SEP systems.

8.0 Conclusions

- SEP has desirable performance potential for several missions.
- Solar cell technology is "in the target range".
- Technology efforts needed to bring thrusters and power processors to TRL 6 at desired power levels and operating life.
- Reasonable design strategies exist for high power/mass and high power systems.
- Major efforts required on lightweighting at component and systems level to reach target 15 kg/kWe or better.
- SEP cost ranges are challenging but not outside space hardware experience.

APPENDIX F

Cost Team Report

IISTP Phase I Final Report
September 14, 2001

Appendix F Cost Analysis Report
Mahmoud Naderi MSFC
Robert Sefcik GRC
Sharon Czarnecki (SAIC)
Gordon Woodcock (Gray Research)

Introduction and Purpose

Cost analyses were performed for missions and technologies to support overall IISTP analysis and scoring. The cost analyses covered only DDTE, hardware acquisition, and launch costs, which were assessed based on current published commercial launch vehicle cost data. Other cost categories considered by IISTP were technology advancement cost and mission operations cost. Operations cost as accounted by IISTP included launch costs, for which we obtained estimates.

Methods

DDT&E and hardware acquisition (unit) costs were estimated using NAFCOM, which is a NASA mass-based parametric (cost estimating relationships) method. NAFCOM has a broad data base of many spacecraft and space vehicles, which can be selectively applied. In this cost analysis, historic spacecraft of similar complexity were used.

We observed that the data base contained two classes of planetary spacecraft of comparable complexity, (1) most of the data base, traditional designs such as Viking and Voyager, and (2) a few spacecraft that represented the "faster, better, cheaper" approach to program management, such as Pathfinder. We chose to bias our selection of spacecraft data base to the traditional cases since it was our understanding that the agency is currently returning to that approach. This caused our estimates to be higher than would be the case if we biased towards the latter sample.

NAFCOM required that we provide estimates of spacecraft mass according to major subsystem. For some of the missions considered, we did not have that information. Therefore, we constructed a representative breakout from the data base and used those percentages to allocate the total mass estimate among subsystems for purposes of the cost estimates.

For the solar electric and nuclear propulsion systems, Bob Sefcik of the Glenn Research Center (GRC) developed and provided the estimates, since NAFCOM has no historical data base for these systems and GRC has applicable experience in systems and technology development.

Missions/Technologies Analyzed

Most of the analysis was concentrated on the Neptune Orbiter since that was the first mission analyzed and scored. NAFCOM runs were made for seven technologies: state-of-the-art chemical, state-of-the-art chemical with aerocapture, advanced chemical, solar electric propulsion (SEP) ion, nuclear electric propulsion (NEP) ion, solar sails, and nuclear thermal propulsion (NTP).

IISTP Phase I Final Report
September 14, 2001

Summary of Results

Two results are presented here: The Cost Team estimates for the Neptune Orbiter mission, and JPL Team X estimates for the Titan Explorer mission. Certain things were unclear about the Team X estimates: (1) no costs for aerobrake could be found in their detailed estimate sheets, so the Cost Team estimates for the Neptune Orbiter aerobrake were added to the Team X estimates for a degree of consistency; (2) it was not clear to what degree the Team X estimates covered DDT&E for in-space propulsion systems. Comparing the estimates, it is suspected that the Team X estimates cover only purchase (unit) costs for these systems.

Neptune Orbiter

	SOA Chem	Chem/AC	Adv Chem	NTP/AC	NEP ion/AC	Solar Sail/AC	SEP/AC
DDT&E							
Main Propulsion	156	152	152	737	537	217	114
Arrival Propulsion	57		75				
Aerobrake		47		47	0	47	47
Unit							
Main Propulsion	42	42	42	42	246	73	89
Arrival Propulsion	19		24				
Aerobrake		10		10	0	10	10
Ops							
Mission Ops TBD							
Launch	172	172	172	111	111	122	172
Total							
Main Propulsion	198	194	194	779	784	290	203
Arrival Propulsion	76		98				
Aerobrake		57		57	0	57	57
Launch	172	172	172	111	111	122	105

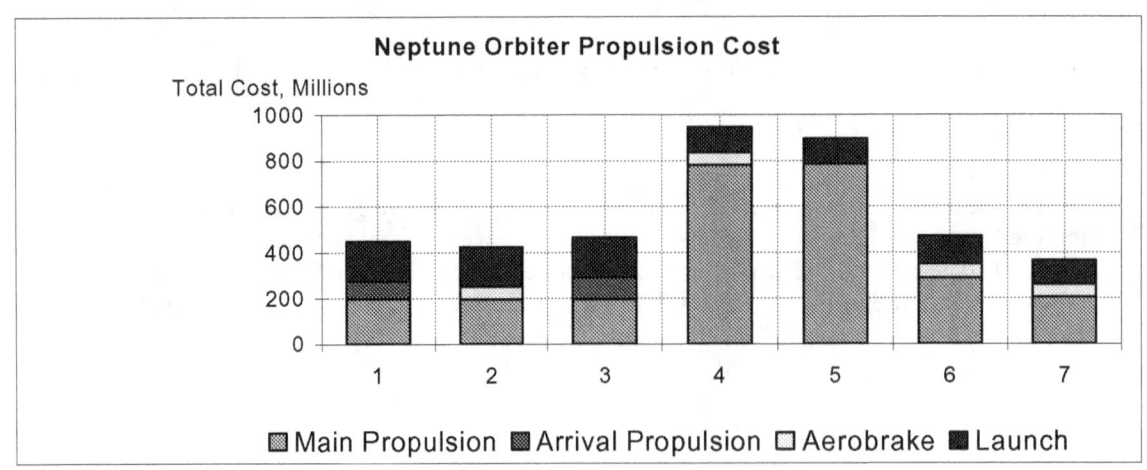

F-4

Neither of the estimates as presented here cover mission operations costs. Team X provided mission operations costs in their estimates, but it does not appear that any unique costs which might be attributed to the technical characteristics of the propulsion systems were identified. Perhaps such unique costs would be too small to be significant. The Cost Team did not estimate mission operations costs.

		Titan Explorer (JPL Team X)						
	Chem/AC	5 kW SEP	10 kW SEP	NEP	NTP	NTP Bimodal	Solar Sail	Plasma Sail
Acquisition								
Main Propulsion	48	143	137	304	140	159	59	87
Arrival Propulsion								
Aerobrake	47	47	47		47		47	47
Launch	114	107	107	107	97	160	97	97
Totals								
Main Propulsion	48	143	137	304	140	159	59	87
Arrival Propulsion								
Aerobrake	47	47	47	0	47	0	47	47
Launch	114	107	107	107	97	160	97	97

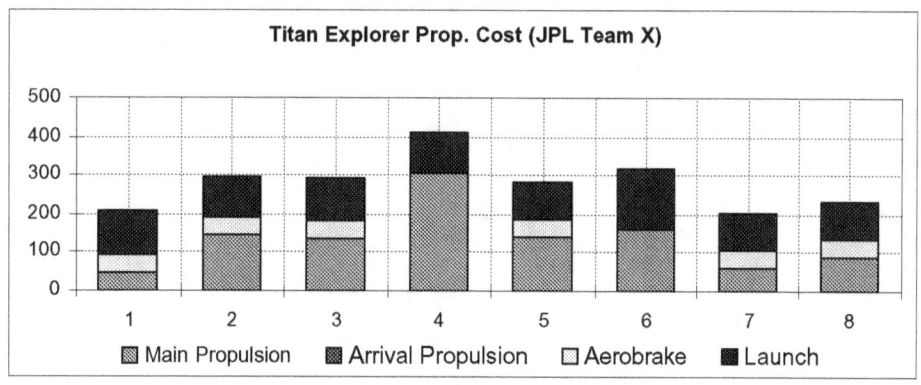

IISTP Phase I Final Report
September 14, 2001

The Cost Team estimates showed nuclear propulsion systems to be considerably more costly than the other systems. Most of this was in the development cost for these systems. The costs are presented as if the Neptune Orbiter program would fund the entire development of the nuclear systems. For major new technology developments this is not realistic. A technology advancement and development program would bring the system to first flight status and the first mission user would pay only the unit cost for the propulsion unit used on the mission. Evaluated in this way, our estimate for nuclear thermal propulsion was not markedly greater than for chemical propulsion; nuclear electric propulsion was still higher in cost than the other systems. Nuclear propulsion costs may not be complete; the NASA estimates may not include reactor costs. The reactors would be developed and produced by the Department of Energy.

It is important to note that while the chemical propulsion systems were competitive in cost they were not competitive in trip time. The chemical/aerobraking option had acceptable trip time.

Costing Ground Rules and Assumptions and Estimator's Notes - Details

General Ground rules

- NASA/Air Force Cost Model (NAFCOM99) was utilized as the primary tool to develop cost estimates for all stages including the propulsion technologies.

- Other costing tools which may provide a better estimate for any specific technology will be used in conjunction with NAFCOM.

- NAFCOM estimates are comprised of Phase C/D costs only.

- Estimates generated for mission concepts include a s/c bus, applicable propulsion technologies, a launch vehicle and propellant.

- Contingency, Program Support and Fee are included in the estimates and are 30%, 15% and 10% respectively.

- Estimates are in Fiscal Year 2001 Dollars (Millions).

- Mass is displayed in kilograms.

- A prototype development approach is assumed for all estimates.

- Systems Test Hardware (STH) quantities used in the model were agreed upon by the Systems Team and are applied to all estimates:
    - STH Qty of 0.25 will be used for all hardware with modular design.
    - STH Qty of 0.5 will be used for all systems using dual identical hardware.
    - STH Qty of 1.0 is used for all other hardware.

- Launch vehicle costs were extracted from AIAA/Isakowitz, 3rd edition (published December 1999) and throughput into the estimates.

- Launch vehicle modification cost estimates are a percentage of the launch vehicle cost.

- Propellant costs are extracted from Standard Prices for Missile Fuels Management Category Item (CMAL NO 00-6) dated 9 August 2000.

Masses for each concept were obtained by GW from the Systems Team except where specifically noted in ground rules of specific concepts. This includes masses for those propulsion technologies estimated by GRC.

Acronyms

AC - Aerocapture
ADCS - Automated data and communications system
BOL - Beginning of life
DCIU - Digital control interface unit
DOE - Department of Energy
FH - Flight hardware
FU - First unit
GRC - Glenn Research Center (formerly Lewis Research Center)
IISTA - In-space Integrated Space Transportation Activity
IISTP - In-space Integrated Space Transportation Planning
Isp - Specific impulse
JPL - Jet Propulsion Laboratory
LH2 - Liquid hydrogen
LO2 - Liquid oxygen
MPD - Magnetoplasmadynamic (MPD) thruster
MSFC - Marshall Space Flight Center
NAFCOM - The cost model, stands for NASA/Air Force Cost Model
NEP - Nuclear electric propulsion
NTP - Nuclear thermal propulsion
PPU - Power processing unit (power conditioning system)
PMAD - Power management and distribution subsystem (PPU is part of the PMAD if not broken out separately)
RTG - Radioisotope thermoelectric generator
SEP - Solar electric propulsion
S/M - Structures and mechanisms
SOA - State of the art
STH - Systems test hardware
TCS - Thermal control system
TNC - Trans-Neptune capture
VaSIMR - Variable specific impulse magnetoplasma rocket
WAG - Rough guess

Initials

GW - Gordon Woodcock (Systems and Cost Teams)
SHC - Sharon Czarnecki (Estimator)

Note: Acronyms not listed are probably spacecraft or space vehicle names.

Launch Vehicles

The following assumptions were concerning Launch Vehicles that will be used for the following missions:

Mission	Launch Vehicles	Comments
Neptune Orbiter Solar Sail w/ AC	Atlas V 550	From IISTP Systems Analysis Team Neptune Orbiter Mission Package, 3/26/01, Larry Kos
Neptune Orbiter Baseline (SEP Ion with AC)	Delta IV Heavy	Direction from Systems Team (GW)
Neptune Orbiter Advanced Chemical	Delta IV Heavy	From IISTP Systems Analysis Team Neptune Orbiter Mission Package, 3/26/01, Larry Kos
Neptune Orbiter State of the Art (SOA) Chemical with Aerocapture	Delta IV Heavy	From IISTP Systems Analysis Team Neptune Orbiter Mission Package, 3/26/01, Larry Kos
Neptune Orbiter State of the Art (SOA) Chemical	Delta IV Heavy	From IISTP Systems Analysis Team Neptune Orbiter Mission Package, 3/26/01, Larry Kos
Neptune Orbiter NEP Ion with Aerocapture	Atlas 530	From IISTP Systems Analysis Team Neptune Orbiter Mission Package, 3/26/01, Larry Kos
Neptune Orbiter NEP VaSIMR with AC	Delta IV M+	From IISTP Systems Analysis Team Neptune Orbiter Mission Package, 3/26/01, Larry Kos
Neptune Orbiter NEP MPD with AC	Assumed Atlas 530	WAG

Ground rules for the Neptune Orbiter S/C Bus

Software was added due to less-than-current amount of software in historic missions

Mass statements were only available at the total level; therefore mass was broken out into the subsystem level by allocating according to subsystems of analogous systems (Galileo Orbiter, Mars Global Surveyor, Mars Observer and NEAR).

STH Quantities are 1.0 for all subsystems

Complexities and Inheritance factors were reviewed and agreed upon or provided by Systems Analysis Team Rep. They are:
D&D and Unit Complexities: 1.0 for all subsystems (at this point the design is not fleshed out enough for anyone to make a judgment call on D&D or Unit Complexity)

Inheritance Factors (see table below).

Inheritance Factors		
Subsystems	Inheritance Factors	Rationale
Structures/ Mechanisms	1.0	New structure & analogous systems are New structure
TCS	0.6	Assume 25% new design. Eng. assumption is that there is significantly more heritage in the thermal subsystem than analogous missions. Existing components but new configuration.
Electrical Power Subsystem	1.0	Radio isotopes are new design and "new generation of generators". Heredity of analogies are discussed below. Galileo Orbiter: New design RTGs (113-0296 pg 3). Mariner 10 & Mariner 6: solar arrays/batteries - reasonably high heredity. NEAR: 4 solar arrays, 1 NiCd battery - assume high heredity was low cost and flew in 96 so solar arrays around for a long time at that point. Pioneer 10 - RTGs some inheritance from earlier Pioneer s/c. Viking Orbiter - solar panels, NiCd. Voyager - RTGs, NiCd supposedly used existing DOE hardware RTGs but high DDT&E to FU ratio so suspect low heredity. Vast majority of the cost is low heredity so I left Inheritance factor at a 1.0.
Power Distribution/Regulation/Control Subsystem	0.5	Assumed 25% new design. Analogy heredity appears to follow that discussed in the Electrical Power Subsystem. Galileo Orbiter: New design (113-0296 pg 3). Mariner 10 & Mariner 6: reasonably high heredity based on other Mariner missions. NEAR - probably high heredity. Pioneer-10 little heredity based on other Pioneer missions but probably not much (based on high DDT&E to FU ratio). Voyager and Viking appears to have very little heredity (high per lb. cost, high DDT&E to FU ratio).
Data Management	0.5	Assumed 25% new design. Analogies heredity: most likely Mars Global Surveyor and NEAR have high heredity; Galileo Orbiter, Mars Observer, Pioneer-10 and Viking Orbiter do not. Vast difference in the lowest and highest datapoints (might want to consider eliminating them). Due to the very high DDT&E cost of Viking Orbiter and the probability that that must have been new design, a lower Inheritance factor is considered appropriate.
Communication	0.7	Assumed 25% new design. Majority of the cost is made up of assumed new design (or significant new design) hardware. Therefore, a lower Inheritance factor was deemed appropriate. (some new design - will need a new antenna as the antenna for Galileo Orbiter did not work!).
ADCS	0.7	Assumed 25% new design. Majority of the cost is made up of assumed new design (or significant new design) hardware. Therefore, a lower Inheritance factor was deemed appropriate.

System analogies were reviewed and agreed upon or provided by Systems Analysis Team Rep. They are:

Analogies

Subsystems	Analogous Missions	Rationale
Structures/ Mechanisms	Outer Planetary Structure/Mechanical Group	Deemed most representative of subsystem (provided by GW)
TCS	Outer Planetary TCS	Deemed most representative of subsystem (provided by GW)
Electrical Power Subsystem	Outer Planetary Components/Subsystems.	Components were chosen because the EPS did not include the anything from the Galileo Orbiter. Selecting "Components" enabled selection of Generator from the Galileo Orbiter. The Pioneer 10 and the Galileo Orbiter had radioisotope. (selected by SHC)
Power Distribution/Regulation/Control Subsystem	Outer Planetary Components/Subsystems.	Components were chosen because the Power Dist/Reg/Ctrl Subsystem did not include anything from the Galileo Orbiter. Selecting "Components" enabled selection of the 3 components that comprise the Power Dist... Galileo Orbiter. (selected by SHC)
Data Management	Outer Planetary Data Management subsystems	Deemed most representative of subsystem (provided by GW)
Communication	Outer Planetary Data Management subsystems	Deemed most representative of subsystem (provided by GW)
ADCS	Outer Planetary Data Management subsystems	Deemed most representative of subsystem (provided by GW)

Ground rules for the Aerocapture

STH quantities are 1.0 for both subsystems.

Complexities and Inheritance factors were reviewed and agreed upon or provided by Systems Analysis Team Rep. They are:
D&D and Unit Complexities: 1.0 for all subsystems (at this point the design is not fleshed out enough for anyone to make a judgment call on D&D or Unit Complexity)
Inheritance Factors (see table below).

Inheritance Factors		
Subsystems	Inheritance Factors	Rationale
Chem Propulsion (Reaction Control)	0.8	Assume some inheritance sense not that different from existing RCS systems (provided by GW)
Aerocapture	1.0	Assume new design (GW). Therefore, Inh factor of 1.0, as Galileo Probe struc also new design. (SHC)

System analogies were reviewed and agreed upon or provided by Systems Analysis Team Rep. They are:

Analogies		
Subsystems	Analogous Missions	Rationale
Chem Propulsion (Reaction Control)	Outer Planetary, Reaction Control, monopropellant filter.	Chosen by SHC
Aerocapture	Galileo Probe Structure	The Galileo Probe aeroshell was recommended as an analogy by GW. The aeroshell is not in the NAFCOM database; therefore, the Struc/Mech subsystem was used as an analogy. NAFCOM description for the Galileo Probe follows: "The Galileo Probe Deceleration Structure is made of aluminum. The main structural member, the aeroshell consists of the payload support ring, a saft box section ring, three longerons connecting the two rings, and the think skin sections. The Module consists of the heat shield, the structure that supports the heat shield, and the parachute subsystem." According to GW, the Neptune Orb Aerocapture will be the Gal. Probe aeroshell plus the heat shield. This must withstand more than other existing heat shields therefore more complex..

Ground rules for SEP Ion

The entire SEP estimate was provided by Bob Sefcik, GRC. All complexity factors, inheritance factors and analogies were chosen by him and included in the estimate he provided.

Per GW, Xenon propellant would be used (used 5.67 kg. per 1000 liters as a conversion factor. 638 kg. of propellant required.).

QNHA		
Subsystem	QNHA	Rationale
Thrusters/Gimbals	4	Assumes four thruster/gimbal sets.
PPUs	4	Assumes four PPUs, two for use and two backup.
DCIU	1.0	
Tanks & Feed System	1.0	
Structure	1.0	
Batteries	1.0	
Ultraflex Solar Array	4.0	Assume four 6 KW BOL Ultraflex arrays.

STH Qtys are provided below:

STH Qty	
Subsystem	STH Quantities
Thrusters/Gimbals	0.25
PPUs	0.25
DCIU	1.0
Tanks & Feed System	1.0
Structure	1.0
Batteries	0.25
Ultraflex Solar Array	0.25

Inheritance Factors for all of the SEP subsystems are 1.0.

D&D and Unit complexities are described below:

D&D and Unit Complexities			
Subsystem	D&D Complexity	Unit Complexity	Rationale
Thrusters/Gimbals	2.5	2.5	Complexity factors were derived to allow for the use of current NAFCOM data for advanced thrusters. The proposed thrusters at 3800 sec Isp require technology development to meet the technology cutoff date of 2005. Recent GRC technology estimates were ~ $4.4M R&D and ~$8.3M full cost.
PPUs	1.0	1.0	
DCIU	1.0	1.0	
Tanks & Feed System	1.0	1.0	
Structure	1.0	1.0	
Batteries	1.0	1.0	
Ultraflex Solar Array	5.0	2.0	DDT&E and FH hardware complexity adjustments required to estimate the impact of designing and building lower weight systems while using a weight-based model.

Analogies follow (no rationale for the choice of analogies was provided):

Analogies	
Subsystem	Analogies
Thrusters/Gimbals	GRO Thruster, Lunar Prospector Thruster
PPUs	Avionics from HETE, Lewis, STEP3, TOMSEP
DCIU	Avionics from HETE, Lewis, STEP3, TOMSEP
Tanks & Feed System	Reaction Control Subsystem from Lewis, Lunar Prospector, Mars Global Surveyor, Mars Pathfinder, NEAR, TOMSEP
Structure	Mars Global Surveyor Structures Subsystem; Mars Pathfinder Cruise Stage Structure
Batteries	Mars Pathfinder Battery
Ultraflex Solar Array	Mars Pathfinder Solar Array

IISTP Phase I Final Report
September 14, 2001

Ground rules for NEP Ion

The entire NEP estimate was provided by Bob Sefcik, GRC. All complexity factors, inheritance factors and analogies were chosen by him and included in the estimate he provided. His note at the beginning of his estimate: "Nuclear Electric Propulsion (NEP) Stage estimate prepared by Bob Sefcik, GRC, using NAFCOM analogies adjusted where needed for advanced technologies. Some throughputs were used where estimating by technical parameters provided a better estimate than estimating by weight."

Didn't know what to use so I assumed same Xenon propellant as used for SEP(used 5.67 kg. per 1000 liters as a conversion factor. 638 kg. of propellant required.).

Bob Sefcik made several references to mass in his notes. These follow:
Thrusters/Gimbals: Used Steve Oleson, GRC, mass estimate of 16.5kg/thruster.
PPUs: PPU was excluded from data provided from MSFC. Added in by Bob Sefcik at a mass of 31.2kg/PPU per Steve Oleson estimate from GRC.
Structure: Structure mass at 10% of total stage = 3592 * .1 = 359 kg

QNHA		
Subsystem	QNHA	Rationale
Thrusters/Gimbals	20	The quantity of twenty was modeled although not deemed to be needed for 100 KWe operation unless the stage was to be reusable.
PPUs	10	
Tanks & Feed System	1.0	
Structure	1.0	
Reactor	1.0	100 KWe SP-100 type
Power Conversion	4.0	4 -25 KWe Brayton
PMAD	4.0	Assumed to be 4 power electronics modules.
Heat Rejection	4.0	Assumed to be 4 modular units.
Heat Exchanger	2.0	Need to check if other heat transport components are needed.

STH Qtys are provided below:

STH Qty		
Subsystem	QNHA	Rationale
Thrusters/Gimbals	0.2	STH quantity set at .2 vs. .25 to arrive at four units for testing with two PPUs.
PPUs	0.2	STH quantity set at .2 vs. .25 to arrive at two full units.
Tanks & Feed System	1	
Structure	1	
Reactor	1	
Power Conversion	1	
PMAD	0.25	
Heat Rejection	1	
Heat Exchanger	1	

Inheritance Factors for all of the SEP subsystems are 1.0.

D&D and Unit complexities are described below:

D&D and Unit Complexities			
Subsystem	D&D Complexity	Unit Complexity	Rationale
Thrusters/Gimbals	2.5	2.5	
PPUs	1.0	1.0	
Tanks & Feed System	1.0	1.0	
Structure	1.0	1.0	
Reactor			
PMAD	1.0	0.5	
Power Conversion			
Heat Rejection			
Heat Exchanger			

Analogies follow (no rationale for the choice of analogies was provided):

Analogies	
Subsystem	Analogies
Thrusters/Gimbals	GRO Thruster, Lunar Prospector Thruster
PPUs	Avionics from HETE, Lewis, STEP3, TOMSEP
Tanks & Feed System	Reaction Control Subsystem from GRO, NEAR
Structure	CRESS Structure; GRO Secondary Structure; TOPEX Structure, Module Support
Reactor	Throughput equation
PMAD	Power Distribution/Regulation/Control Subsystem for Lunar Prospector, Mars Pathfinder, NEAR
Power Conversion	Throughput equation
Heat Rejection	Throughput equation
Heat Exchanger	Throughput equation

IISTP Phase I Final Report
September 14, 2001

Ground rules for SOA Chemical

Mass statements were provided at the component level by GW.

STH Quantities are 1.0 for all subsystems

FIRST STAGE (Trans Neptune Insertion TNI LOX/LH2 Stage)

Complexities and Inheritance factors provided by GW with one exception. They are: D&D and Unit Complexities: 1.0 for all subsystems (at this point the design is not fleshed out enough for anyone to make a judgment call on D&D or Unit Complexity). However, the Pressurization System has D&D and Unit complexities of 0.5 to compensate for the use of a manned analogy. (SHC)

Inheritance Factors (see table below).

Inheritance Factors		
Subsystems	Inheritance Factors	Rationale
RL-10B-2		Throughput cost of RL-10A-3-1 from NAFCOM
Fwd skirt	1.0	
Thr str/AS	0.5	
Tanks	0.3	Assume buy off-the-shelf and modify slightly (appx 20% new design). Standard tanks (not composite tanks) per GW
Tank Insulation	1.0	
Intertank	1.0	
Avionics	1.0	
Pressurization System	0.5	All complexities set at 0.5 because of the use of the older data set (manned analogy from Apollo LM)
Feed	0.5	

Analogies were selected by SHC and are provided below:

Subsystems	Analogies	Rationale
RL-10B-2		Throughput cost of RL-10A-3-1 from NAFCOM
Fwd skirt	Galileo Orbiter, Structure; Mars Global Surveyor Structures Subsystem; Mars Pathfinder Cruise Stage Structure	
Thr str/AS	Galileo Orbiter, Structure; Mars Global Surveyor Structures Subsystem; Mars Pathfinder Cruise Stage Structure	
Tanks	Tanks for DSP, GPSMYP, GRO, HEAO-1, HEAO-2, HEAO-3, Lunar Prospector, Pioneer Venus Bus/Orbiter, TOPEX	Filtered on Tanks for Unmanned and chose all missions that had Tanks resulting from the search
Tank Insulation	Thermal Control Subsystem for Centaur D, Centaur G', S-IC, S-II, S-IVB	Estimated differently than the Stor. prop. stage because of the cryogenic tanks. Therefore, analogies are Liquid Launch Vehicle Stages Thermal Control subsystems (much of the cost was due to Insulation Blankets around the LH2 tanks).
Intertank	Tanks for DSP, GPSMYP, GRO, HEAO-1, HEAO-2, HEAO-3, Lunar Prospector, Pioneer Venus Bus/Orbiter, TOPEX	Team decision to use same analogies as used for Tanks.
Avionics	Avionics from 70 missions that had Avionics in the Unmanned Earth Orbiting Database	Avionics for stage should be relatively simple complexity.
Pressurization System	Pressurant Components from Apollo LM	Only pressurization system I could find broken out into component level in the database.
Feed	DSCS-II Feed Components; Lunar Prospector Lines, Valves, Filters	

SECOND STAGE (Trans Neptune Capture TNC Stage)

This is a bipropellant stage (per GW). I assumed monomethylhydrazine and nitrogen tetroxide (per SHC WAG).

Complexities and Inheritance Factors used are the same as that used in the FIRST STAGE for each subsystem present in SECOND STAGE. Complexities and Inheritance factors provided by GW with one exception. They are:
D&D and Unit Complexities: 1.0 for all subsystems (at this point the design is not fleshed out enough for anyone to make a judgment call on D&D or Unit Complexity). However, the Pressurization System has D&D and Unit complexities of 0.5 to compensate for the use of a manned analogy. (SHC)

Inheritance Factors (see table below).

Inheritance Factors		
Subsystems	Inheritance Factors	Rationale
Thr str/AS	0.5	
Tanks	0.3	Assume buy off-the-shelf and modify slightly (appx 20% new design). Standard tanks (not composite tanks) per GW
Tank Insulation	1.0	
Pressurization System	0.5	All complexities set at 0.5 because of the use of the older data set (manned analogy from Apollo LM)
Feed	0.5	
Thrusters	1.0	
Avionics	1.0	

Analogies were selected by SHC and are provided below:

Subsystems	Analogies	Rationale
Thr str/AS	Galileo Orbiter, Structure; Mars Global Surveyor Structures Subsystem; Mars Pathfinder Cruise Stage Structure	
Tanks	Tanks for DSP, GPSMYP, GRO, HEAO-1, HEAO-2, HEAO-3, Lunar Prospector, Pioneer Venus Bus/Orbiter, TOPEX	Filtered on Tanks for Unmanned and chose all missions that had Tanks resulting from the search
Tank Insulation	Tank Insulation: ATS-6, DSP, INTELSAT-IV, Mars Pathfinder, OMV, UFO	Used different analogies than for LOX-LH2 insulation as it should be simpler (no cryogenic tanks). Analogies are a mix of Unmanned Earth Orbiting and Planetary blanket components.
Pressurization System	Pressurant Components from Apollo LM	Only pressurization system I could find broken out into component level in the database.
Feed	DSCS-II Feed Components; Lunar Prospector Lines, Valves, Filters	
Thrusters	Thrusters: DSP, GPSMYP, GRO, HEAO-1, HEAO-2, HEAO-3, Lunar Prospector, Pioneer Venus Bus/Orbiter	Unmanned EO and Planetary Thrusters (selected all that had Thrusters broken out)
Avionics	Avionics from 70 missions that had Avionics in the Unmanned Earth Orbiting Database	Avionics for stage should be relatively simple complexity.

Ground rules for Solar Sail

Solar sail: 10 g/cm2 sail at 350,000 m2

Mass statements were provided at the component level by GW on the Monster Weights Rev 1 spreadsheet.

STH Quantities are 1.0 for all subsystems

Complexities were assumed to be 1.0.

Inheritance Factors were selected by SHC and provided below:

Inheritance Factors		
Subsystems	Inheritance Factors	Rationale
Unique S/M	1.0	New structure but all structures are new therefore no adjustment necessary to factor.
Repeat S/M	1.0	Same as above
Sail Membrane	1.0	New material, process, fabrication etc. Could make a case for increasing inheritance due to new material however I did not because the analogous insulation blankets - although not new material - most of cost is in new design and labor. Also mass much higher than for blankets so cost differential captured here.
Avionics	1.5	Issue with solar sails is "control of large, flexible, lightweight space structures and development of an effective attitude and articulation control system." (Advanced Propulsion Concepts, JPL). Increased avionics heritage to 1.5 because this avionics has not been developed yet ("no operational solar sail tests of yet").

Analogies were selected by SHC (with some guidance by GW) and are provided below:

Subsystems	Analogies	Rationale
Unique S/M	Chose composite Structures/Mechanical Group, Unmanned missions as analogous datapoints.	Assumed that much of Unique S/M was the deployment mechanism therefore made cost per kg. higher than for the Repeat S/M.
Repeat S/M	Chose composite Structures subsystem, Unmanned missions as analogous datapoints.	Assumed that much of Repeatable S/M was mainly comprised of structure for each of the four segments of the sails (some mechanisms such as gimbals) - and therefore made cost per kg. lower than for the Unique S/M.
Sail Membrane	Chose Unmanned missions, Insulation Blankets as analogous datapoints.	Sail membrane is comprised of a single layer plastic film (aluminized). GW suggested that multilayer insulation would be the best analogy on a cost per lb. basis. The sail membrane will probably be manufactured in strips then joined together not unlike MLI so labor cost similar.
Avionics	Used all Earth Orbiting Avionics subsystem missions for analogy.	Avionics are new but relatively simple. Just used for controlling gimbals? not sending and receiving signals?

IISTP Phase I Final Report
September 14, 2001

Ground rules for NEP MPD

I have not received input from Bob Sefcik, GRC, at this time. Therefore, I copied the NEP Ion estimate that he had generated and used that as a basis for MPD. Per Cost Team decision, I left the masses unchanged for NEP Ion although they differed significantly from those provided in the Monster Weight Rev 1 spreadsheet. Therefore, in the below discussion when stated that a subsystem is "unchanged" it will not match the Monster Weight spreadsheet but the NEP Ion NAFCOM estimate generated by Bob Sefcik.

Per Monster Weight Rev 1 spreadsheet, the power generation subsystems were all the same mass so I left them alone (Reactor, Power conversion, Heat Rejection, Heat Exchanger, PMAD). Structure did not change either so I left it unchanged. Avionics was listed in all three NEP cases in the Monster Wt spreadsheet (NEP Ion, MPD and VaSIMR), but I did not include it in either MPD or VaSIMR as it was not included in the NAFCOM basecase. Tanks mass was changed as this mass changed between subsystems. Thrusters and PPUs were zeroed out until I receive inputs from GRC.

Propellant is xenon; mass is 4400 kg. Or 9700 lbs.

Numerous issues with having the mass in my NAFCOM model differ from that given in Monster Weights. DISCUSSION POINT - I strongly feel they should agree.

QNHA		
Subsystem	QNHA	Rationale
Thrusters/Gimbals	2+2	From Gordon's email, 2 thrusters plus 2 spares. 50 kg. Apiece. (Should this weigh 200 kg. Rather than 240 kg?) Does this include gimbals and such??
PPUs	2	From Monster Spreadsheet - would there be any spares here? Do they weig 250 kg. Together (125 kg.s apiece) or does each weigh 250 kg.?
Tanks & Feed System	1.0	
Structure	1.0	
Reactor	1.0	100 KWe (from Gordon's email) SP-100 type (assume same as NEP Ion case)
Power Conversion	4.0	ASSUMED SAME AS NEP ION CASE 4 -25 KWe Brayton
PMAD	4.0	ASSUMED SAME AS NEP ION CASE Assumed to be 4 power electronics modules.
Heat Rejection	4.0	ASSUMED SAME AS NEP ION CASE Assumed to be 4 modular units.
Heat Exchanger	2.0	ASSUMED SAME NEP ION CASE Need to check if other heat transport components are needed.

STH Qtys are provided below: ASSUME SAME AS NEP ION

STH Qty		
Subsystem	QNHA	Rationale
Thrusters/Gimbals	0.2	STH quantity set at .2 vs. .25 to arrive at four units for testing with two PPUs.
PPUs	0.2	STH quantity set at .2 vs. .25 to arrive at two full units.
Tanks & Feed System	1	
Structure	1	
Reactor	1	
Power Conversion	1	
PMAD	0.25	
Heat Rejection	1	
Heat Exchanger	1	

Inheritance Factors for all of the SEP subsystems are 1.0. ASSUME SAME AS NEP ION

D&D and Unit complexities are described below:

D&D and Unit Complexities			
Subsystem	D&D Complexity	Unit Complexity	Rationale
Thrusters/Gimbals	Sefcik	Sefcik	Waiting for Sefcik data
PPUs	Sefcik	Sefcik	Waiting for Sefcik data
Tanks & Feed System	1.0	1.0	
Structure	1.0	1.0	
Reactor			
PMAD	1.0	0.5	
Power Conversion			
Heat Rejection			
Heat Exchanger			

IISTP Phase I Final Report
September 14, 2001

Analogies follow (no rationale for the choice of analogies was provided): EXCEPTING THRUSTER AND PPU - ASSUME SAME AS NEP ION.

Analogies	
Subsystem	Analogies
Thrusters/Gimbals	Awaiting Sefcik data
PPUs	Awaiting Sefcik data
Tanks & Feed System	Reaction Control Subsystem from GRO, NEAR
Structure	CRESS Structure; GRO Secondary Structure; TOPEX Structure, Module Support
Reactor	Throughput equation
PMAD	Power Distribution/Regulation/Control Subsystem for Lunar Prospector, Mars Pathfinder, NEAR
Power Conversion	Throughput equation
Heat Rejection	Throughput equation
Heat Exchanger	Throughput equation

IISTP Phase I Final Report
September 14, 2001

Ground rules for NEP VaSIMR

I have not received input from Bob Sefcik, GRC, at this time. Therefore, I copied the NEP Ion estimate that he had generated and used that as a basis for VaSIMR. Per Cost Team decision, I left the masses unchanged for NEP Ion although they differed significantly from those provided in the Monster Weight Rev 1 spreadsheet. Therefore, in the below discussion when stated that a subsystem is "unchanged" it will not match the Monster Weight spreadsheet but the NEP Ion NAFCOM estimate generated by Bob Sefcik.

Per Monster Weight spreadsheet, the power generation subsystems were all the same mass so I left them alone (Reactor, Power conversion, Heat Rejection, Heat Exchanger, PMAD). Structure did not change either so I left it unchanged. Avionics was listed in all three NEP cases in the Monster Wt spreadsheet (NEP Ion, MPD and VaSIMR), but I did not include it in either MPD or VaSIMR as it was not included in the NAFCOM basecase. Tanks mass was changed as this mass changed between subsystems. Thrusters and PPUs were zeroed out until I receive inputs from GRC.

Propellant is liquid hydrogen, mass is 4403 kg. Or 9707 lbs.

Numerous issues with having the mass in my NAFCOM model differ from that given in Monster Weights. DISCUSSION POINT - I strongly feel they should agree.

QNHA		
Subsystem	QNHA	Rationale
Thrusters/Gimbals	2+2	From Gordon's email, 2 thrusters plus 2 spares. 50 kg. Apiece. (Should this weigh 200 kg. Rather than 240 kg?) Does this include gimbals and such??
PPUs	2	From Monster Spreadsheet - would there be any spares here? Do they weig 250 kg. Together (125 kg.s apiece) or does each weigh 250 kg.?
Tanks & Feed System	1.0	
Structure	1.0	
Reactor	1.0	100 KWe (from Gordon's email) SP-100 type (assume same as NEP Ion case)
Power Conversion	4.0	ASSUMED SAME AS NEP ION CASE 4 -25 KWe Brayton
PMAD	4.0	ASSUMED SAME AS NEP ION CASE Assumed to be 4 power electronics modules.
Heat Rejection	4.0	ASSUMED SAME AS NEP ION CASE Assumed to be 4 modular units.
Heat Exchanger	2.0	ASSUMED SAME NEP ION CASE Need to check if other heat transport components are needed.

STH Qtys are provided below: ASSUME SAME AS NEP ION

STH Qty		
Subsystem	QNHA	Rationale
Thrusters/Gimbals	0.2	STH quantity set at .2 vs. .25 to arrive at four units for testing with two PPUs.
PPUs	0.2	STH quantity set at .2 vs. .25 to arrive at two full units.
Tanks & Feed System	1	
Structure	1	
Reactor	1	
Power Conversion	1	
PMAD	0.25	
Heat Rejection	1	
Heat Exchanger	1	

Inheritance Factors for all of the NEP subsystems are 1.0. ASSUME SAME AS NEP ION

D&D and Unit complexities are described below:

D&D and Unit Complexities			
Subsystem	D&D Complexity	Unit Complexity	Rationale
Thrusters/Gimbals	Sefcik	Sefcik	Waiting for Sefcik data
PPUs	Sefcik	Sefcik	Waiting for Sefcik data
Tanks & Feed System	1.0	1.0	
Structure	1.0	1.0	
Reactor			
PMAD	1.0	0.5	
Power Conversion			
Heat Rejection			
Heat Exchanger			

Analogies follow (no rationale for the choice of analogies was provided): EXCEPTING THRUSTER AND PPU - ASSUME SAME AS NEP ION.

Analogies	
Subsystem	Analogies
Thrusters/Gimbals	Awaiting Sefcik data
PPUs	Awaiting Sefcik data
Tanks & Feed System	Reaction Control Subsystem from GRO, NEAR
Cryocooler	Reaction Control Subsystem from GRO, NEAR
Structure	CRESS Structure; GRO Secondary Structure; TOPEX Structure, Module Support
Reactor	Throughput equation
PMAD	Power Distribution/Regulation/Control Subsystem for Lunar Prospector, Mars Pathfinder, NEAR
Power Conversion	Throughput equation
Heat Rejection	Throughput equation
Heat Exchanger	Throughput equation

IISTP Phase I Final Report
September 14, 2001

Ground rules for SEP NSTAR

I have not received input from Bob Sefcik, GRC, at this time. Therefore, I copied the SEP Baseline estimate that he had generated and used that as a basis for SEP NSTAR. Per Cost Team decision, I left the masses unchanged for SEP Baseline although they differed significantly from those provided in the Monster Weight Rev 1 spreadsheet. Therefore, in the below discussion when stated that a subsystem is "unchanged" it will not match the Monster Weight spreadsheet but the SEP Baseline NAFCOM estimate generated by Bob Sefcik.

Thrusters/Gimbals and PPU costs were zeroed because I have no input from GRC at this time. All other analogies and complexity factors for every subsystem are the same for NSTAR as for the SEP Baseline. The only changes made were to masses for Structures/Mechanisms and Tanks & Feed System. Propellant load was also adjusted.

Per GW, Xenon propellant would be used (used 5.67 kg. per 1000 liters as a conversion factor. 470 kg. of propellant required. Per Monster Weights Rev 1 spreadsheet).

ASSUMED SAME QNHA AS SEP BASELINE

QNHA		
Subsystem	QNHA	Rationale
Thrusters/Gimbals	4	Assumes four thruster/gimbal sets.
PPUs	4	Assumes four PPUs, two for use and two backup.
DCIU	1.0	
Tanks & Feed System	1.0	
Structure	1.0	
Batteries	1.0	
Ultraflex Solar Array	4.0	Assume four 6 KW BOL Ultraflex arrays.

STH Qtys are provided below: **ASSUMED SAME STH QTY AS SEP BASELINE**

STH Qty	
Subsystem	STH Quantities
Thrusters/Gimbals	.25
PPUs	.25
DCIU	1.0
Tanks & Feed System	1.0
Structure	1.0

Batteries	0.25
Ultraflex Solar Array	0.25

Inheritance Factors for all of the SEP subsystems are 1.0. ASSUMED SAME INHERITANCE AS SEP BASELINE.

D&D and Unit complexities are described below: ASSUMED SAME COMPLEXITY FACTORS AS SEP BASELINE.

D&D and Unit Complexities

Subsystem	D&D Complexity	Unit Complexity	Rationale
Thrusters/Gimbals	2.5	2.5	Complexity factors were derived to allow for the use of current NAFCOM data for advanced thrusters. The proposed thrusters at 3800 sec Isp require technology development to meet the technology cutoff date of 2005. Recent GRC technology estimates were ~ $4.4M R&D and ~$8.3M full cost.
PPUs	1.0	1.0	
DCIU	1.0	1.0	
Tanks & Feed System	1.0	1.0	
Structure	1.0	1.0	
Batteries	1.0	1.0	
Ultraflex Solar Array	5.0	2.0	DDT&E and FH hardware complexity adjustments required to estimate the impact of designing and building lower weight systems while using a weight-based model.

Analogies follow (no rationale for the choice of analogies was provided): ASSUMED SAME ANALOGIES AS SEP BASELINE.

Analogies

Subsystem	Analogies
Thrusters/Gimbals	GRO Thruster, Lunar Prospector Thruster
PPUs	Avionics from HETE, Lewis, STEP3, TOMSEP
DCIU	Avionics from HETE, Lewis, STEP3, TOMSEP
Tanks & Feed System	Reaction Control Subsystem from Lewis, Lunar Prospector, Mars Global Surveyor, Mars Pathfinder, NEAR, TOMSEP
Structure	Mars Global Surveyor Structures Subsystem; Mars Pathfinder Cruise Stage Structure
Batteries	Mars Pathfinder Battery
Ultraflex Solar Array	Mars Pathfinder Solar Array

www.ingramcontent.com/pod-product-compliance
Lightning Source LLC
Chambersburg PA
CBHW081719170526
45167CB00009B/3636